DECOLONIZING EXTINCTION

EXPERIMENTAL FUTURES
Technological Lives, Scientific Arts, Anthropological Voices
A series edited by Michael M. J. Fischer and Joseph Dumit

DECOLONIZING EXTINCTION

THE WORK OF CARE IN ORANGUTAN REHABILITATION

Juno Salazar Parreñas

DUKE UNIVERSITY PRESS DURHAM AND LONDON 2018

Printed in the United States of America on acid-free paper ∞
Designed by Heather Hensley
Typeset in Garamond Premier Pro by Copperline Book Services

Library of Congress Cataloging-in-Publication Data
Names: Parreñas , Juno Salazar, [date–] author.
Title: Decolonizing extinction : the work of care in orangutan rehabilitation / Juno Salazar Parreñas .
Description: Durham : Duke University Press, 2018. | Series: Experimental futures : technological lives, scientific arts, anthropological voices | Includes bibliographical references and index.
Identifiers: LCCN 2018004044 (print) | LCCN 2018007841 (ebook)
ISBN 9780822371946 (ebook)
ISBN 9780822370628 (hardcover : alk. paper)
ISBN 9780822370772 (pbk. : alk. paper)
Subjects: LCSH: Orangutans—Borneo. | Wildlife rehabilitation—Borneo. | Human-animal relationships.
Classification: LCC QL737.P94 (ebook) | LCC QL737.P94 P37 2018 (print) | DDC 599.88/3095983—dc23
LC record available at https://lccn.loc.gov/2018004044

Cover art: Photo by Juno Salazar Parreñas

To Lakas Parreñas Shimizu,
Mary Margaret Steedly,
and the person I call Layang

CONTENTS

ACKNOWLEDGMENTS

This book is dedicated to the person I call Layang and to my nephew Lakas. Their untimely deaths and larger-than-life personalities have occupied my thoughts since 2011 and 2013. My PhD advisor Mary Margaret Steedly passed away as this book was in production. I could not have accomplished this project without her steadfast support and inspiring wisdom. May they rest in peace.

I remain indebted to the many people who let me into their lives and workplaces while conducting this research, from September 2008 to July 2010, in Sarawak, Malaysia. Many conversations and shared moments deeply influenced my ideas in this book, and I cannot properly thank many of my interlocutors here, for fear that they will be blamed as individuals for greater systemic problems. In particular, I thank a person who is not mentioned in the book, but who explained to me that police in Sarawak are armed like the military of other countries, and who then asked me, "It's worse in the Philippines, isn't it?" I also thank the friend whom I call Nadim, whose humor and kindness served as a buffer from what is ultimately a painful job. The depth of empathy he has is truly inspiring. The person who I call Cindy is the personification of joy, wisdom, and hilarity, and I am eternally grateful for her introducing me to fresh coconut-sugar cane juice. If the world had more people like Nadim and Cindy, we would all be in a better place.

Dr. Charles Leh and Director Ipoi Datan of the Sarawak Museum; Professor Datuk Dr. Abdul Rashid Abdullah of the Institute of East Asian Studies, the International Affairs Division of the Universiti Malaysia Sarawak; Sarawak's State Planning Unit of the Chief Minister's Department; Meena S. Ponnusamy and Dr. James Coffman of the Malaysian-American Commission on Educational Exchange, as well as Ngui Siew Kong and Engkamat Lading of the Forestry Department graciously hosted my research while I was in Sarawak. I benefited greatly from the conversations taking place at UNIMAS's IEAS, now known as the Borneo Studies Center, in particular with Kelvin Egay and Jayl Langub. I especially thank Kelvin for his friendship and enthusiasm for ethnographic methods. I value the time Barbara Harrisson and G. S. Silva gave me.

I cannot sufficiently thank the people of the longhouse in which I lived for their hearty welcome and gifts of conviviality, time, stories, experiences, and good humor. I have very deep gratitude to Ing, a super-friend and sister by choice. I am grateful to Philip, who surely withstood neighborhood gossip in accommodating me in Kuching, and to Rachel for letting me stay with her husband! I thank Philip's parents for hosting me during Chinese New Year, too, and for all the delicious mangosteen, durian, and moon cake. I thank Steph, Mikey, Florence, and Christine for their cheerfulness, warmth, conversations, good times, and excellent taste in food. I am grateful to Leykun for welcoming me to the gang.

A Fulbright–IIE fellowship sponsored a year of this research. In graduate school at Harvard University, the Cora Du Bois Summer Fellowship and Graduate Society Fellowship supported my writing. Foreign Language and Area Studies Fellowships at the University of Wisconsin at Madison and the University of California at Berkeley, as well as a Harvard University Asia Center language grant sponsored my language training. Grants from the Office of International Affairs and an Arts and Humanities Small Grant at The Ohio State University enabled the completion of this book.

Mary Steedly, in her role as advisor and committee chair, gave me both the support and the freedom to develop my writing and research. She consistently believed in this project. I am grateful for her mentorship, kindness, and fun honesty. Steve Caton went above and beyond the expectations of being what he had jokingly called a "warm seat." This project would have been impossible without Donna Haraway. Her challenging questions and enduring support since 2006 have nourished and encouraged me to continue to stay with the trouble. I benefited greatly from conversations with and feedback from Amitav Ghosh, Ajantha Subramanian, Engseng Ho, Smita Lahiri, Byron Good, Mike Fischer, Michael Herzfeld, Pauline Peters, Jill Constantino, Karen Strassler, Eduardo Kohn, Nancy Peluso, Cori Hayden, Tom Lacquer, and James Vernon.

I learned much from my peers and friends during my time at Harvard, UC Berkeley, and SEASSI, in particular Darryl Li, Ujala Dhaka, Will Day, Namita Dharia, Nico Cisterna, Sabrina Peric, Julie Kleinman, Naor Ben-Yehoyada, Dadi Darmadi, Kedron Thomas, Pete Benson, Kian Goh, Tamiko Beyer, Andrew Littlejohn, Rusaslina (Lin) Idrus, Lindsay Smith, Miriam Shakow, Lilith Mahmud, Anthony Shenoda, Maryann Shenoda, Amali Ibrahim, Veronika Kusumaryati, Vernie Oliviero, Illiana Quimbaya, Andie Murray, Clare Gillis, Felicity Aulino, Adia Benton, Andrea Allen, Jesse Grayman, Fumi Wakamatsu, Bernardo Zacka, Alireza Doostdar, Alex Fattal, Claudio Sopranzetti, Maria Stalford, Alexandra Daiferro, Lyndon Gill, Emily Hammer, Parker VanValkenburgh, Tina Warinner, Janice Calleja, Ana Huang, Sarah Rodriguez, Jia Hui Lee, Luis Campos, Emilie L'Hote, Sandra Bruenken, Katie Hendy, Sarah Grant, Meredith Root-Bernstein, Natalie Porter, and Chika Watanabe. I also thank others involved with Harvard University's Political Ecology Working Group and Southeast Asia Workshop for fruitful and inspiring discussions. Amy Zug, Susan Farley, Susan Hilditch, and Marianne Fritz were always wonderfully helpful, even across great distances. I remain grateful to Lindsay Smith and her encouraging me in 2004 to pursue the project that started to grab my attention away from my original intentions.

The Animal Studies Institute and Human-Animal Studies summer fellowship at Wesleyan University fostered multispecies conviviality with Lori Gruen, Kari Weil, Kēhaulani Kauanui, Gunnar Eggertsson, Brigitte Fielder, Sarah Hann, Robert Jones, Eliza Ruiz Izaguirre, and Harlan Weaver.

A postdoctoral fellowship at Yale University in the Agrarian Studies Program gave me the unrestrained time to dive into Yale's Southeast Asian library sources to develop chapters 4 and 5. I thank Kalyanakrishnan Sivaramakrishnan and Jim Scott for fostering that special intellectual community. I am especially grateful for Shivi's generous mentorship over the years. It was a pleasure sowing seeds of knowledge and perennial friendship with Jami Mukherjee, Gabe Rosenberg, Matthew Bender, Todd Holmes, Atreyee Majumder, and Radhika Govindrajan.

A postdoctoral fellowship at Rutgers Center for Historical Analysis for a seminar on "Networks of Exchange, Mobilities of Knowledge" with historians Toby Jones and James Delbourgo allowed me to join Rutgers's dynamic intellectual community. I thank Nicole Fleetwood for welcoming me to the Institute for Research on Women's seminar on "Decolonizing Gender, Gendering Decolonization," which was a theme proposed by Yolanda Martinez-San Miguel. It was a sincere pleasure to think with Toby Jones, James Delbourgo, Nicole Fleetwood, David Hughes, Temma Kaplan, Julie Livingston, Ann Fabian, Chie Ikeya, Judith Surkis, Seth Koven, Angelique Haugerud, Preetha Mani, Sarah Tobias, Nadia Guessous, Ghassan Moussawi, Annie Fukushima, Nova Robinson, Marian Thorpe, and Erin Vogel. My deepest gratitude goes to Lynn Shanko, Toby Jones, Temma Kaplan, and Nicole Fleetwood.

At The Ohio State University, I thank my former chairs Jill Bystydzienski and Guisela Latorre, and current chair Shannon Winnubst for their support of this book project. I thank Lynaya Elliot for her organizational talents. I am grateful for Tess Pugsley and her artwork. Comments from Cricket Keating, Shannon Winnubst, Noah Tamarkin, Jenny Suchland, Mytheli Sreenivas, Joe Ponce, Lynn Itagaki, Mary Thomas, Max Woodworth, Brian Rotman, Katherine Marino, Nick Kawa, Dodie McDow, Melissa Curley, Inéz Valdez, Lisa Bhungalia, Adam Thomas, Monamie Bhadra, Daniel Rivers, Ben McKean, Dana Howard, Becky Mansfield, and especially Danilyn Rutherford improved this book. I also thank my current and former colleagues in the Women's, Gender, and Sexuality Studies Department at The Ohio State University: Judy Wu, Wendy Smooth, Treva Lindsay, Cricket Keating, Shannon Winnubst, Jenny Suchland, Mytheli Sreenivas, Lynn Itagaki, Mary Thomas, Guisela Latorre, Azita Ranjbar, Linda Mizajewski, and Katherine Marino. I am grateful for the joyful and generous intellectual community of the Space and Sovereignty working group, in particular past and present members Noah Tamarkin, Becky Mansfield, Dodie McDow, Melissa Curley, Katherine Marino, Sarah van Beurden, Nada Moumtaz, Lisa Bhungalia, Adam Thomas, and Monamie Bhadra.

I benefited greatly from the comments and questions I received when I presented different iterations of this book at York University's Faculty of Environmental Studies, Indiana University's School of Global and International Studies, Cornell University's Ronald and Janette Gatty Lecture Series, University of Chicago's Anthropology Department, the Tri-College Symposium of Women of Color in Academia at Bryn Mawr College, the Comparative History of Ideas Program at the University of Washington, the University of Chicago's Working Groups on Animal/Nonhuman Studies and Comparative Behavioral Biologies, the Nanyang Technical University of Singapore Sociology Department, the University of Malaya Gender Studies Program, the University of Texas Austin's Anthropology Department, New York University's Anthropology Department, Natalie Porter's seminar at Notre Dame, and Tim Choy's graduate seminar at UC Davis. These visits were brief but meaningful for me. Thanks in particular to Stacy Rosenbaum, Sam Schulte, and Zoe Hughes for the hands-on feedback and generative discussion about chapter 3. I am especially grateful for a quick conversation with Joe Masco about time. I thank Radhika Govindrajan and Jayedev Athreya for their hospitality. At the University of Malaya, I thank Rusaslina Idrus for facilitating my visit and allowing me to see the amazing community that she is fostering with Associate Professor Dr. Shanthi Thambiah and their colleagues.

I have been extremely lucky to participate in workshops where we had the luxury of time and resources to think together. An Intimate Industries workshop at Pomona College organized by Hung Thai, Rhacel Parreñas, and Rachel Silvey fostered exciting discussions with Ju Hui Judy Han, Akhil Gupta, Purnima Mankekar, Daisy Deompampo, Sharmila Rudrappa, and others. The Decolonizing Science in Asia seminar at Penn State allowed me to workshop chapter 6 with a sharp and generous group of Asianists: Prakash Kumar, Amit Prasad, Projit Mukherjee, Bharat Venkat, Dwai Banerjee, Burton Cleetus, Charu Singh, Lan Li, Nicole Barnes, Quentin Pearson, Gabriela Soto Laveaga, and Lijing Jiang. The How Nature Works seminar at the School for Advanced Research influenced my thinking about time, labor, and plantations thanks to Sarah Besky, Alex Blanchette, Naisargi Dave, Thomas Andrews, Eleana Kim, Shiho Satsuka, Jake Kosek, Kregg Harrington, Al Nading, Maria Elena Garcia, and John Hartigan. The working group on captivity fostered by Kevin O'Neill at the University of Toronto and Jatin Dua at the University of Michigan has been a source of inspiration. I thank them for fostering the opportunity to think with them, as well as with Darryl Li, Noah Tamarkin, Andrew Shyrock, and Susan Lepselter. A seminar on Volatile Futures/Earthly Matters at the Center for the Advancement of Public Action, hosted by David Bond and Joe Masco, fostered discussions with Alex Blanchette, Andrea Ballestro, Amy Thomas Moran, Lucas Basiere, Kristina Lyons, and Nick Shapiro. A seminar at the Max Planck Institute for the History of Science, on Decolonizing the Plan II, organized by Kavita Phillip, Anidita Nag, Martina Schluender, Emily Brownwell, Sarah Blacker, and Sarah van Bluerden, allowed me to think more about decolonization. Feedback at the University of Amsterdam from Annemarie Mol, Amade M'charek, Maria Fernanda Olarte-Sierra, Rebeca Ibáñez Martín, Emily Yates-Doerr, Carolina Dominguez-Guzman, Lisette Jong, Ildiko Plajas, and others at a workshop organized by Emily and Carolina that was cosponsored by the RACEFACEID and Eating Bodies research teams inspired me to rethink my hunches. I especially thank Amade M'charek for the short residency at UVA and welcoming me to her dynamic research team. Working with sound artists Nicolas Perret and Silvia Ploner (Island Songs) was an inspiration.

It was both a pleasure and a learning experience to work with zookeeper Cathy Keyes at the Oakland Zoo in the summer of 2008. I also thank Gail Campbell-Smith, Panut Hadsiswoyo, and Ian Singleton. Thanks to Kelly Boyer Ontl and Andrew Halloran for inviting me to the International Primatology Society and the American Society for Primatology joint conference in Chicago, to Chris Schmitt for sharing his love for monkeys, to Ryan Cadiz for introducing me to Chris, and to graduate students from Erin Vogel's lab Shauhin Alavi, Tim Bransford, and Alysse Moldawer who let me hang out with them at Lincoln Zoo. I thank the various participants at IPS ASP 2016 who conveyed their excitement about this project. I hope it isn't disappointing!

I am deeply grateful to those who gave me written feedback on the manuscript before it went into publication, especially Danilyn Rutherford, Joe Klein, and the two anonymous reviewers. I am very thankful for opportunities to think through key ideas with Jayum Jawan, Shivi, and Bob Root-Bernstein. A version of chapter 2 appeared in *American Ethnologist*, and parts of chapter 4 appeared in *Positions: Asia Critique*. I thank both journals for their peer review process. Thanks to Linda Forman, Christine Byl, and David Heath for their copyedits. The smooth journey into production is all thanks to Ken Wissoker, Elizabeth Ault, and Susan Albury at Duke University Press.

As an undergraduate at UC Santa Cruz, Anna Tsing, Carla Freccero, Neferti Tadiar, Wendy Brown, Teresa de Lauretis, Jerry Miller, and Gayle Rubin all had profound impact on my thinking. I would not have been able to finish my undergraduate education without the help of my financial aid adviser Liz Martin-Garcia. UCSC's intellectual kinship has hosted many stimulating conversations with Noah Tamarkin, Harlan Weaver, Danny Solomon, Heather Swanson, Eva Hayward, Martha Kenney, Eben Kirksey, Cindy Rose Bello, Nick Mitchell, Bettina Stoetzer, Sandra Koelle, Jake Metcalf, Logan Walker, Kristina Valendinova, Sarah Baaker, Sarah Kelman, and Colin Hoag.

I deeply appreciate friendships that taught me a lot about intimacy and chosen family in the different cities in which I have lived in the past. I especially thank Ing, Erica Cho, Chi-Ming Yang, Derrick, Luka, Verena, Erdmute, Birgit, Gela, Amanda, Sherryn, Anne, Five Star, Susan, Kat, Lisa, Jane, Gabrielle, Dvora, Chase, Jay, Jessica, Andrea, and Manuela. I cannot thank them enough.

As the third sister with a PhD, I am deeply grateful for my two eldest sisters and their copious and wise advice about everything. Rhacel Parreñas was the one I called when I needed advice about how to write a paper longer than two pages—which was the length I was accustomed to as a public school kid from San Bernardino. Celine Shimizu and I read and talked about Foucault together in the 1990s, often while I helped myself to her pantry. I thank my siblings Rolf, Celine, Rhacel, Rhanee, Cerissa, Mahal, and Aari for their support, humor, advice, and thoughtfulness over the years. My parents; Dan, Sharon, Ian, Claudio, and Ben; Lauren and Ciara, Bayan and Lakas, Javi and Brady, Mica and Caleb; Susan, Marshall, Kenny, Jean, Emily, Tanya, Scott, Michael, Joshua, and Theodore; and Babushka, Pangga, Gugma, Jackie, and especially Mpho all taught me ideas of care, love, and intensity for which I will always be thankful. And finally, I thank Noah, my favorite nocturnal primate, for thinking with me at all hours of the day and night. All shortcomings in this book are mine.

Introduction

DECOLONIZING EXTINCTION

Eight-year-old Lisbet was born in captivity at Lundu Wildlife Center in Sarawak on Borneo. Workers, officers, and volunteers alike commonly spoke of her as an orphan, but there was another story about her early life. The rumor was that when Lisbet was still an infant, her mother had been sent to a resort on a small man-made island in peninsular Malaysia to begin a rehabilitation program for semi-wild orangutans. The last time orangutans freely roamed the peninsula was during the Pleistocene.[1] Orangutans are otherwise only native to Sumatra and Borneo, islands where they found refuge during climate change millions of years ago and survived—unlike their conspecifics who died out on the peninsula and elsewhere in Southeast Asia.[2]

The newly privatized branch of the state's forestry department surely welcomed the source of revenue Lisbet's mother would bring through the memorandum of understanding between the resort and the department. Privatization of the newly forged semigovernmental agency meant the same work as before, but with less money and fewer staff members. They were still responsible for the care of indigenous and endangered wildlife

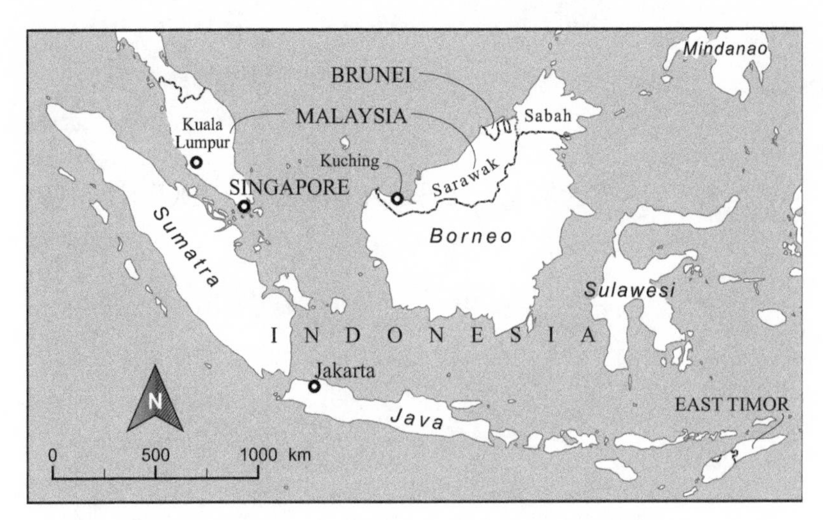

MAP I.1 Malaysia and Borneo in Southeast Asia. Map by James DeGrand.

in Sarawak, a semiautonomous state of Malaysia that is often treated as an internal colony. Lisbet lost her mother to the resort when she was still dependent on her. Yet the people involved felt that the resort could offer better conditions than the Forestry Corporation could. The fact that the resort could afford to keep a veterinarian on staff, for instance, while Lundu Wildlife Center could not, is one of the small ways that make up bigger ways in which colonialism is an ongoing process in Sarawak.

Eight-year-old Lisbet's world was populated with humans. When I first encountered her, she spent stretches of time in her enclosure standing erect on her limbs, so that her hand-like feet curled to support her entire weight. It was a feat considering how her body was adapted to living in trees rather than on the ground. For months, our encounters consisted of me jotting notes in her presence, either in the night house, where iron bars mediated our shared space, or outside, where I stood on a viewing platform two flights of stairs above her enclosure. Lisbet crossed these barriers and made our shared interface more eventful by throwing projectiles at me. Once it was a watermelon rind, other times stones, or her spit. Her eyes would arrest my gaze during such moments. Our social relation was subtle yet significant, at least for me. Orangutans have the reputation of being the most solitary of all the great apes. But the orangutans held at Lundu Wildlife Center, like Lisbet, are neither wild nor tame. The center aspires to teach these orangutans to become semi-wild.[3]

Lisbet's initial experiences outside of captivity occurred in our first week of meeting, and it appeared she was going to fail rehabilitation. Her keepers took her deep into the 22-km^2 forest that surrounded the wildlife center and immersed her in a verdant expanse. Yet Lisbet did not climb a single tree. Instead, she stayed on the ground with the group of men trying to train her.

Layang, one of these men, explained to me what would happen if Lisbet failed rehabilitation. If she were unable to demonstrate such skills as tree climbing or nest building, she would be assigned to the captive breeding program. Once sexually mature at about fourteen years old, she would be temporarily confined with a male orangutan for the purpose of impregnation. Such confinement does not take "female choice" into account. A few months after birth, the infant would be taken away, even though orangutan infants usually stay with their mothers for about seven years, learning how to survive (Galdikas and Wood 1990). Keepers like Layang doubted that a mother whose life has been spent in captivity would be able to teach her infant how to live. It is precisely for this reason that they would then take the infant away from her because they figured they would have a better chance at training the infant in jungle skills than the orangutan's mother.[4]

This is what the future would hold if Lisbet were deemed unable to be rehabilitated. Layang did not want that to happen. He knew the violence Lena experienced in the week she was confined with Efran.[5] That horror was for naught since both Lena and her baby died shortly after Lena gave birth. Layang felt it was worth the shared effort of bringing Lisbet 10 km into the forest, even if someone had to carry her piggyback, and even if she weighed nearly as much as Layang.

Inspired by the burdens Layang and his coworkers endured, my intention in this book is to urge reflection: What if we experienced this present era of extinction without violent domination and colonization over others, particularly nonhuman beings? Can we instead embrace the vulnerability of sharing our lives together, however fleeting those moments might be? Can we abandon an impression of safety that depends on cruelty? In other words, how might we decolonize extinction?[6]

To become semi-wild meant achieving the goal of becoming *bebas*, or the freedom of unrestrained license. This is not the freedom of the postcolonial nation-state, officially celebrated as national holidays (*merdeka* in Malay) (Steedly 2013). Nor is it the freedom of movement espoused in the philosophy of liberalism, where such freedom is limited to the rights of fully fledged citizens (Khalili 2013). Neither is it the libertine's freedom of wild parties, wild nightlife, and wild animals (*liar* in Malay). Bebas is the freedom of acquittal, the independence of factory women unyoking from their fathers' orders (Ong 1987), and the liberation of youth in Indonesia and peninsular Malaysia (Ibrahim 2018; Idrus 2016; Lee 2016). This kind of freedom has no directive goal or a priori destination—or, if it does, it is open to possibilities, uncertainties, and experimentations. As I will show in this work, the sense of liberation and independence in bebas offers a theory of decolonization.

Teaching Lisbet was not an act of domination, and any animal keeper would agree.[7] Working with semi-wild orangutans entailed taking personal risk and experiencing physical vulnerability.[8] Thorny durians that can make humans bleed are one of their favorite foods (Reddy 2012). Their teeth are well suited for chewing bark off trees. Already Lisbet was strong enough to inflict pain by biting flesh, something that semi-wild orangutans are apt to do. Unlike wild orangutans, who try to keep their distance from people, semi-wild orangutans are habituated to people and do not fear them.

Layang felt the risk was worth it. In his opinion, risk, vulnerability, and interest were essential characteristics of the work of care in orangutan rehabilitation.[9] A simple need for a job would be insufficient motivation for this kind of work. Construction jobs or overseas logging and oil rigging work could readily be had. Those other jobs might very well require degrees of risk and vulnerability, but working at the wildlife center meant fostering new and extraordinary kinds of social relations, relations at the very interface of the serious threat of a species's annihilation.[10]

Looking at Sarawak's two wildlife centers hosting semi-wild orangutans, we see how colonial legacies and postcolonial institutions impact the way orangutans live and die. Batu Wildlife Center is situated in a state nature reserve that is 6.5 km^2. When a master's student conducted research there in 1995, it was 15 km^2 (Chow 1996). The normal range for an individual female orangutan is 7 km^2 (Galdikas 1988). This area is meant to accommodate an ever-growing population of orangutans: twenty-six as of 2010.[11]

The reserve is on the edge of the capital city. To the west, sand excavation by the state's largest development firm, Global Limited, cleared the forest on the other side of the boundary.[12] The sand that was extracted from this site was used to build a world-class airport that could better facilitate in-

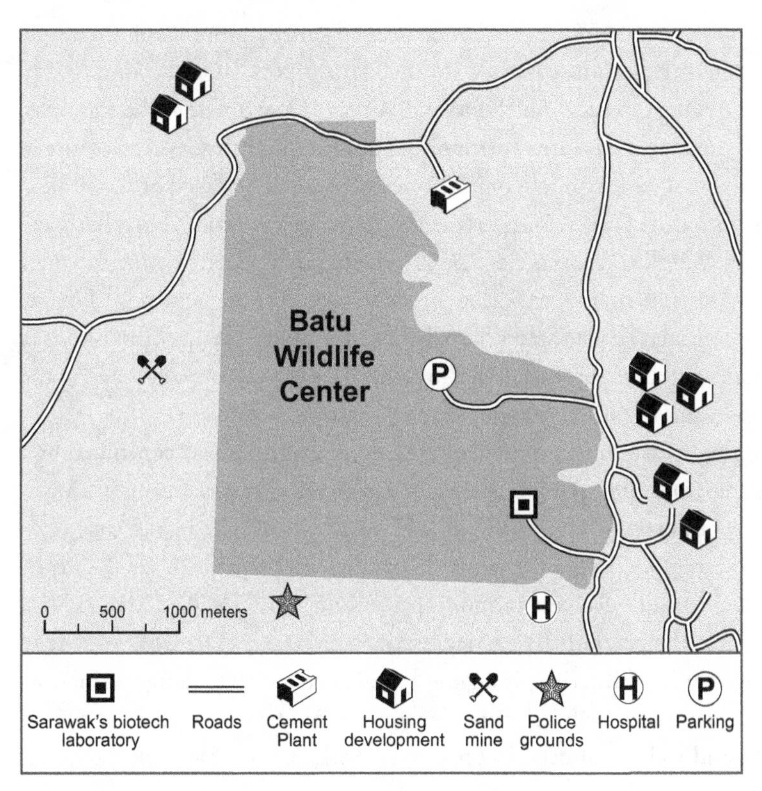

MAP I.2 Batu Wildlife Center. Map by James DeGrand.

ternational and regional commerce, including tourism.[13] To the northwest, simple Malay, Chinese, and Melanau homes and gardens push along the wire fencing of the reserve boundaries. At the northeast boundary, trucks come in and out of a cement factory. To the east, gated housing developments with names like "Borneo Gardens" have sprouted along both sides of the road. Toward the southwest is the police training academy, where practice gunshots in quick succession are heard regularly. To the south lies a hospital. The park manager has repeatedly complained about the pollution caused by the health facility: discarded syringes float by on the creek weaving its way through the nature reserve and out by the simple homes northwest of the site where people fish.

However, when visitors from abroad stand in the middle of the center's courtyard, all they seem to see is forest. As one German visitor said, with awe in her voice, "Is all of Sarawak like this?" Her guide didn't understand her, so she elaborated, "With so many trees?"

In 1997, Lundu Wildlife Center opened on the grounds of a national forest in order to house Batu Wildlife Center's population beyond the latter's carrying capacity. The Land and Survey Department of the state classified the map of the area surrounding Lundu Wildlife Center because of the high risk of illegal logging. The newer site was modeled on Australian zoos. Since the Forestry Department's partial privatization, both Batu and Lundu Wildlife Centers operate as private-public partnerships run by a "semigovernmental" agency known as the Forestry Corporation.[14] The site hosts commercial volunteers, mostly British women, who pay thousands of American dollars per week to perform hard labor that supports Layang and others' work of care in orangutan rehabilitation.

In the institutions created by past colonial regimes and continued by a postcolonial state, we see a theory of decolonization generating in the everyday but extraordinary work of care that happens here in Sarawak's two wildlife centers. To see the work of Layang with Lisbet is to see the experimental work of decolonization in present-day postcolonial Sarawak. It is to realize the possibilities carried by the word *bebas*. The work of care in orangutan rehabilitation, I suggest, is an effort at decolonizing extinction. Care is not necessarily affection, but for me it is a concern about the treatment and welfare of others. This takes work; it takes labor that requires compensation.[15]

Decolonization is not a past era of the mid-twentieth century, ushered in by anticolonial "mimic men" of the postcolony (Bhabha 1994; Mbembe 2001; Wilder 2015).[16] Rather, decolonization is an ongoing process in Sarawak that simultaneously experiences an ongoing colonialism. The stakes of decolonization are not limited to issues of sovereignty, occupation, or knowledge production—all of which are contemporary struggles in decolonization and in continued colonialism more broadly (Allen and Jobson 2016; Fanon 1965; Harrison 1991; Smith 1999; TallBear 2013). Instead, decolonization scratches at fundamental ways of understanding the world.[17] Taking decolonization seriously would entail not just questioning who manages Sarawak's ecologies and how they manage them, as political ecologists have long pushed us to consider (Brosius 1999; Cooke 2006; Dove 2011; Dove et al. 2005; Padoch 1982; Padoch and Peluso 1996; Peluso 1991; Peluso and Lund 2011; West 2006). It also entails questioning deep-seated assumptions about life and ecology: who is living, in what ways are we in

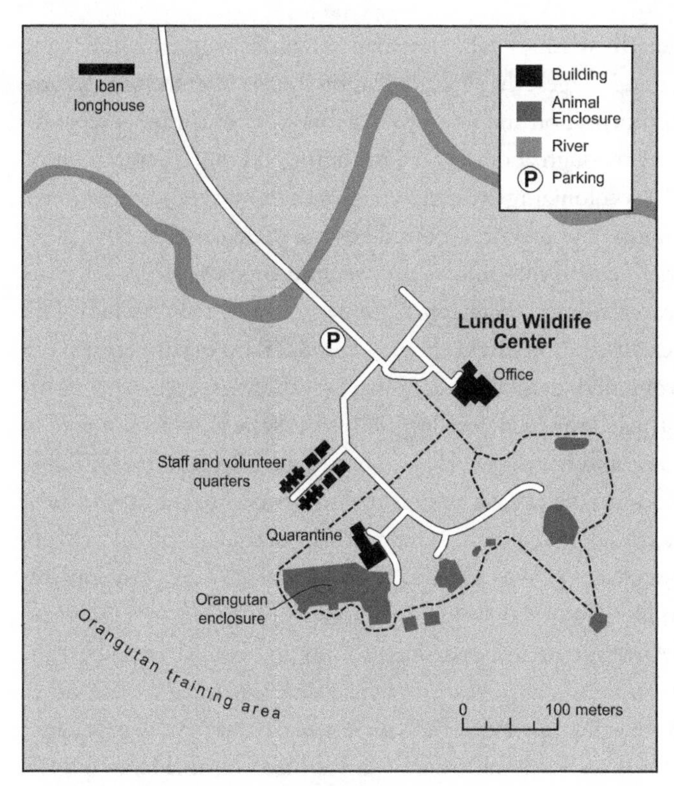

MAP I.3 Lundu Wildlife Center. Map by James DeGrand.

relation with them, what constitutes selves in these relations, and to what obligations are we committed (de la Cadena 2010; Kohn 2013)?[18]

Even as decolonization demands a serious challenge to the so-called great divides between human and animal or inanimate, it also demands a rejection of a telos (Haraway 1991; Latour 1993).[19] To decolonize extinction is to resist definitively saying what should be or ought to be.[20] Indeed, what might look like liberation, such as the free mobility of orangutans within the constraints of a wildlife center, may on a deeper level be less liberatory than it seems. Yet what makes such an action a potential form of decolonization is its experimentation in how to relate to others beyond tired colonial tropes of violence and benevolence.[21]

To seriously consider the impact of our actions on those nearly at the brink of extinction, we need to think about what other ways things might be done, especially when we take the perspectives of orangutans and workers into account. I suggest that decolonization is to be oriented toward process and experimentation and not toward a foregone conclusion, except for the need to care enough about others, including and in particular nonhuman others.[22] Decolonizing extinction requires a serious reconsideration of the current norms and practices around how we share this planet.

The stories I share in this book occur over four timescales. First, they are about affective encounters that happen over seconds and microseconds. This is felt between all kinds of earthly bodies. Second, these occur between different kinds of individuals, whether human or otherwise, each carrying life histories that span years and decades. This is the scale at which we tend to feel space, place, and memory (Feld and Basso 1996; Rosaldo 1980). Third, the connections taking place here must be understood in the *longue durée* that entails a consideration of multiple centuries of trade, mobility, and colonialism (Braudel 1958). We sense this on the spatial scale of oceans and seas (Gilroy 1993; Ho 2006; McDow 2018; Sharpe 2016; Spyer 2000; Subramanian 2009). And fourth, extinction has us thinking about the epochal time of thousands and millions of years, in which time is marked by death on a mass scale. Such a divine perspective is impossible for humans to experience directly, and we can barely touch on it in fossilized form (Haraway 1988; Shryock et al. 2011). It is only apparent to us through the detritus of material bodies that comprise the layers of geologic time (Andrews 2008).[23]

Thinking through these multitudes of timescales simultaneously, we can start to imagine that a single timescale alone is insufficient for understanding the fleeting intimacies that cross many kinds of difference and that happen at such sites as Lundu Wildlife Center. Most importantly, grasping these layers of time frames together points to contingency. By contingency, I mean this: things have not always been the way they are and thus do not have to be this way in the future.

The contingency of our present circumstances frames the central question that guides this book: How are we to live and die in this present age of extinction, when colonial legacies help determine who and what is in better position to survive? Layang, the wildlife ranger Nadim, and the junior officer Cindy offer inspiration for how we can think about and live with the relations that make up the planet as we know it.

Extinction in this book is not a muse for a eulogy about creatures that one nostalgically misses even while actively killing them (Choy 2011; Heise

2016; Rosaldo 1989). If we were to take on an earthbound perspective of multiscalar time, we would see that extinction, like individual death, is a condition of planetary living. Decolonizing extinction is not an attempt to try to stop it. Rather, the question and challenge of decolonizing extinction is its experimentation with other responses and other senses of responsibility than what usually inspires us when we want to do something—anything—to stop what might be inevitable. The challenge of decolonizing extinction, then, is not to end extinction, but to consider how else might it unfold for those who will perish and for those who will survive.

Decolonization appears to be emerging from a frustration with our current moment, whether we call that moment late capitalism, late liberalism, or the Anthropocene. Such terms cut across temporal scales and seem disconnected, but they are indeed inextricably connected. Critical questions such as which bodies—land, human, and otherwise—bear the toxins of industries indict environmental racism and ongoing colonialism, especially settler colonialism and its subtle and not-so-subtle forms of genocide (Bohme 2015; Goeman and Denetdale 2009; Haraway 2016; Murphy 2016). Decolonization emphasizes the politics underlying the ontological turn, which has been accused of being apolitical (Bessire and Bond 2014; Kohn 2013).

Some criticize the emergence of decolonization in scholarship on the grounds that leftist and progressive scholars are simply using the term as a synonym for social justice at large. Doing so loses its specificity and erases its political possibilities, as Eve Tuck and Wayne Yang (2012) argue.[24] In other instances, decolonization gets folded into decoloniality, which cannot be done without abusing the past.[25] While decolonial scholars like Maria Lugones (2010) offer a means to recognize nonhuman others as colonized subjects, decolonial efforts to center colonial exploitation beginning in 1492 and the subsequent sixteenth century ultimately work as a modified world systems theory (Mignolo 2015; Quijano 1995; Wallerstein 1974).[26] To consider world systems theory at this moment, whether called as such or by a new name, suggests that colonialism is singular and far more totalizing and absolute in its power.[27] If we were to accept the hegemony and totality of colonialism, we could not sufficiently consider the possibilities for how things might be otherwise.[28]

My hope is that stories from Sarawak can inspire our aspirations elsewhere for an otherwise, one that does not impose isolation, firmly bounded categories, nor exclusionary nativism, but instead invites a recognition of interdependencies across kinds and differences (Cattelino 2008; Kauanui 2008). The aspiration of decolonization that I perceived in Sarawakians' work of care for Sarawak's wildlife differs from decolonization based on autochthony (Geschiere 2009). Indigeneity in Sarawak, with its more than thirty indigenous ethnic groups, rejects *Blut und Boden*, a German idiom of ethnic nationalism that has resonated in various historical eras and all too easily has led to genocide, ethnic violence, partition, political misrecognition, and forced exile (Mamdani 1996, 2001; Tamarkin 2011). Rather, indigeneity in Sarawak is based on centuries and millennia of migration, both within Borneo and across seas. When we consider deep history and epochal time, we see that indigeneity in Borneo is about mobility and refuge.[29]

This book is not a story of settler colonialism and the ways it kills and dies. It is instead a story of both extractive and internal colonialism generating relations, enclosures, and futures.[30] It is also a story of finding refuge: Ibans migrating within Borneo and in Sarawak to gain a living, orangutans who found refuge from climate change that occurred millions of years ago, and their contemporary descendants who now find refuge in the outskirts of the city of Kuching (Arora et al. 2010).[31] It is also a story of colonization, such as the Sarawakian bacterium *Burkholderia pseudomallei* that I describe later in this book (Podin et al. 2014). The microbe uniquely evolved in Sarawak to live in Sarawakian soil and feed off Sarawakian plants and animals. It is also a story of decolonization, of the work of challenging the impetus for extraction that impacts the orangutans, plants, and people of Sarawak. And, last but not least, it is a story about the politics of extinction, one that is feminist in its commitment to understanding how gender, sexuality, and social inequalities shape how we live through and respond to the threat of species loss, and one that is critical of the colonial legacies that underlie our relations with nonhuman others. When we look at relations in different kinds of timescales and in different and unexpected kinds of spatial formations, we can get a perspective that shows how things that seem so entrenched may not be as permanent or indefinite as they seem.

Extinction

Mass extinctions mark the transition of epochs. Extinction, along with the fluke of mutation, generates coevolution (Cassidy and Mullin 2007; Sodikoff 2012). Earth at the end of the Cretaceous Period 66 million years ago witnessed around 93 percent of terrestrial species dying out (Longrich et al. 2016). They appear in the sediments of rock as traces of lives lost long ago. Recovery is thought to have been fast in geologic time, at 300,000 years (Longrich et al. 2016).

A timescale in which more than a quarter million of years means speedy recovery forces us to recognize that our lives are short, that our moments together are always fleeting. From such a vantage point, we get the sense that we bodily forms have significant impacts on each other until the shells of our mutual existence eventually get embedded in layers of earth. Such a theory of evolution is attached to the rock of this planet and raises the question: How shall we each make our mark?

The Cretaceous–Tertiary extinction event is one of six periods of mass extinction in Earth's history. In the time since then, Borneo has become a "biodiversity hotspot" of mythic proportion and a native habitat for an array of endangered species like the three subspecies of Bornean orangutan (Mittermeier et al. 1998; Myers and Mittermeier 2000). In the ecological turmoil of the Pleistocene three million years ago, orangutans found refuge from fluctuating ice ages (Arora et al. 2010).[32] Borneo, site of the world's oldest rain forest, continues to foster the coevolution of new life forms.

As many of us already know, we are currently in another moment of extinction. Asteroids, volcanoes, or meteors did not begin this current wave of destruction. The sixth extinction marks the end of the Holocene and the beginning of the Anthropocene. The current wave of extinction is thought to be pushed by the homogenization of flora and fauna, the high proportion of biomass consumed and then wasted by humans, the heavy hand humans have exercised on certain domesticated animals, and the technosphere of roads, power plants, and the taken-for-granted comforts of modernity (Williams et al. 2015). Picture for a moment rows of mono-crops, cattle feed lots with waste runoff, and the trucks and cars on the highway passing them all by: that is what extinction looks like. A quarter of mammals on Earth are threatened, endangered, or critically endangered. Orangutans are one of many that are now fewer than ever before.

The extinction of our epoch bears a moral weight. The response to mass extinction has not been to curb the burning of fossil fuels or to cease the standardization of species in industrial agriculture.[33] However, one acute response has been to directly intervene in the lives of endangered species. The moral weight of extinction is significant enough to generate an indus-

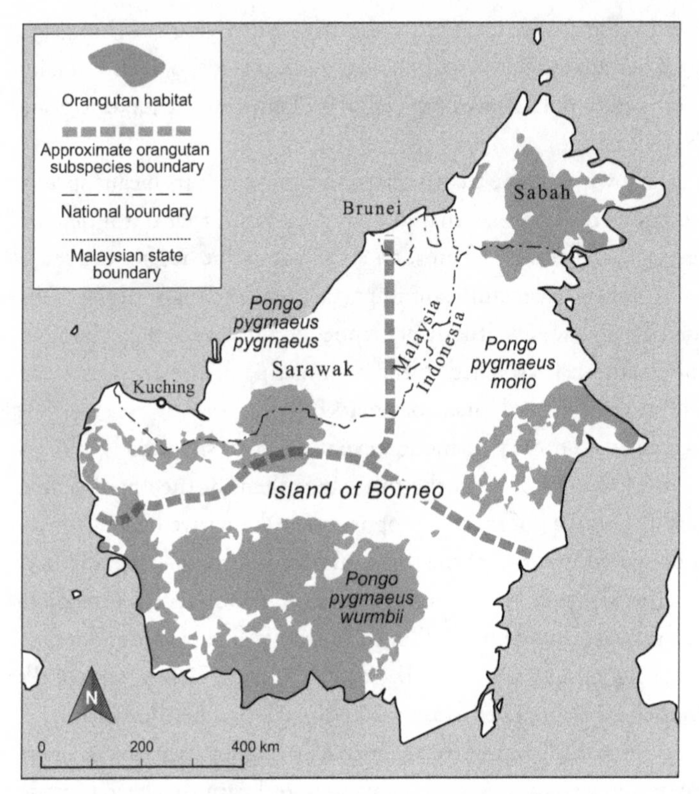

MAP 1.4 Bornean Orangutan Subspecies Population Distribution. Map by James DeGrand, based on data from Wich and Kuehl (2016) and the *IUCN Red List of Threatened Species*, http://www.iucnredlist.org, downloaded on May 21, 2017.

try of volunteer tourism for threatened wildlife. Individuals with the financial resources to participate in commercial volunteerism can personally feel part of a greater mission. The "mission" in which they engage is secular. Commercial volunteerism for endangered wildlife lacks the language of religious ideology in its goals or motivations. Commercial volunteers come to Lundu Wildlife Center not for salvation, but motivated by a professed interest in animals and conservation.

When describing the plight of orangutans—and all orangutan subspecies are now critically endangered as of 2016—primatologists and conservationists sometimes emphasize that habitat loss has an extreme impact on orangutans because of their long birth intervals and low reproduction rates (Cawthon Lang 2005). In other words, the blame for extinction falls partially on sexual reproduction. Thus female orangutans in particular bear the burden of their survival as a species. We get a hint of this when we consider the designation of "captive for breeding purposes." A similar hint appeared when a former CEO of the Forestry Corporation boastfully and impossibly promised that the state of Sarawak would target an increase in the population of orangutans to nearly double its present size (Chan 2009).

The International Union for Conservation of Nature serves as the authority on extinction, and they periodically assess the status of endangered species. Their 2016 assessment, published as part of their continuously updated Red List of Threatened Species, explains that the blame for orangutans' status as critically endangered lies in two factors: the destruction, degradation, and fragmentation of their habitats and hunting. The argument that hunting is a noteworthy cause of extinction relies on a quantitative survey (Meijaard et al. 2011). That study found that cases in which people killed orangutans had stemmed from conflicts that arose when orangutans raided farmers' crops.[34] While the study shows that crop raiding by orangutans was more common at sites that were surrounded by monocultural industrial agriculture, particularly palm oil, rice, pulp, and paper plantations, the blame is nevertheless attributed to peasants' hunting practices instead of agricultural industrialization (Voight et al. 2018).

When scientists feel that the conservation of orangutans lies in the hands of powerful "decision makers" and not the people who directly interface with the species in question, then indigenous hunters of Borneo and female orangutans slow to get pregnant become easy scapegoats for the problem of extinction (Meijaard et al. 2012; Wich and Kuehl 2016). My work in these pages offers an alternative view by highlighting the perspectives and experiences of those who are often blamed for orangutan extinction: displaced female orangutans and the people on the ground who work with them, including displaced indigenous people.

The workers coming from the Iban longhouse outside Lundu Wildlife Center are displaced. The orangutans Ching and Ti hail from Batang Ai, the very place from which the caretaker and *Tuai Rumah* (Iban longhouse headman) Apai Julai came. It is not a coincidence, since Batang Ai then was the site of a large hydroelectric dam construction project that affected a water catchment of more than 1,200 km^2 (Cramb 1979; King 1986; Sarawak Museum 1979). This and other examples show that both wildlife and people in Sarawak are subject to ecological loss, and such loss creates new social relations across species.

When viewing human–orangutan relations with four simultaneous time scales, we look to the future as well as the past. Over hundreds, thousands, and millions of years, we get glimpses of adaptation and resilience. When we think of the future, can we do so without a response to mass death that depends on sexual reproduction and the rearing of younger generations? Can we expand our imaginations to envision other ways of living and dying at the temporal and spatial brink of extinction?[35] Can we, like the wildlife ranger Nadim, make serious efforts to "think what the orangutan are thinking"?

Orangutans

Tourists come to Sarawak's wildlife centers explicitly to encounter orangutans. When they do, they encounter a variety of other inhabitants, including endemic trees, flowering plants, squirrels, and bats. Visitors originally fixated on orangutans find themselves captivated by other species, such as gibbons, sun bears, macaques, and binturong. These wildlife centers stress particular interest in orangutans, and in doing so they suggest a hierarchy of species, which I personally find difficult to espouse. My purpose here is not to argue that orangutans should take priority over other life forms. I want to think about the relations that develop between orangutans and the people who care for them as examples of how we inhabit this planet with others in the current age of extinction.

We cannot know with certainty an orangutan's perspective. Even if we were able to follow the synapses of orangutans' neurons, we still wouldn't know what it *feels* like to be an orangutan. A sense of who an orangutan is or what she might become is limited to signs conveyed by their bodies. Some of these signs may not even be perceivable to you or me. We could turn to different kinds of experts to help us piece together what orangutan perspectives might be.

A behavioral ecologist might tell you that orangutans are known as the least social of all hominids, which is not the same as thing lacking social relations. Birth intervals among orangutans are the longest of all the great apes, with a seven- to eight-year gap the typical average between pregnancies (Galdikas and Wood 1990; Kuze et al. 2008). Their lives are semi--

solitary: they tend to live and travel independently if they are not part of a mother–infant dyad, a temporary group of juvenile males, or a temporary coupling.

A conservationist, on the other hand, will likely tell you that orangutans, like other great apes, can be divided into wild, captive, and rehabilitant populations. Wild populations of orangutans, a conservationist will privately admit, are the most important. They are the ones who live in biodiverse habitats. They are the ones who serve as a "flagship species," with efforts toward their preservation saving large swaths of forest, and with it other, less charismatic creatures (Barua 2011; Root-Bernstein et al. 2013). They are the ones whose behaviors are more unknown and thus more interesting.

Captive populations of orangutans are now mostly zoo animals, since bans on the use of apes in medical research began across the world in the early 2000s (Knight 2008; Nihon Kankyō Kaigi 2009). When survival in the wild is tenuous, captivity potentially becomes the sole means of survival for a species (Braverman 2015). Yet "extinct in the wild," an official designation by the International Union for Conservation of Nature, is a mere step before flat-out "extinct."

For our hypothetical conservationist, rehabilitation poses a problem. Rehabilitation centers give sanctuary to displaced apes who were often caught in illegal trafficking and taken into custody by the state. Some conservationists criticize sanctuaries for promoting genetic admixture, at times across subspecies, which is a problem for those who value gene pools with diversity (Goossens et al. 2009). Rehabilitant orangutans have been exposed to human contact and anthroponotic illnesses, which endanger wild populations. This is why primatologist Herman Rijksen abandoned rehabilitation in the 1970s. Rehabilitation is also considered more expensive than other conservation operations when dollar figures are calculated per individual ape (Meijaard et al. 2012).

Great ape rehabilitation centers are tourist sites, and although nearly every facility in the world has explicit recommendations about the proper distance to protect against respiratory illness, tourists nevertheless often show up at these sites with symptoms of illnesses that can harm the very endangered apes they came from afar to see (Muehlenbein et al. 2010). It should not surprise anyone that rehabilitation centers have high infant mortality rates, higher than either wild or captive populations (Kuze et al. 2012).

These different primatological perspectives can help us get a sense of what a pubescent, twelve-year old female orangutan might feel as the yearning for solitude in the midst of a group of twenty-six others packed into a forest that would accommodate only one orangutan in the wild. Yet privileging a primatologist's perspective over all others limits our imagination to those with technoscientific expertise (Haraway 1988). Adding more perspectives widens our scope and offers vantage points we may not have had the sense to notice.

Scholars of human–animal studies use stories and experiences to evoke feelings that are actively suppressed in most other contemporaneous science writing. Think of the loving human hands laid on reputedly bright laboratory rats (Despret 2004), the encouragement whispered to fighting crickets (Raffles 2010), or the frustration of Indonesian primatologists struggling and failing to have a Sulawesi macaque recognized as a distinct species (Lowe 2006). Multispecies ethnographers in particular value embodied ways of knowing (Dave 2014; Govindrajan 2015; Hayward 2010; Parreñas 2012, 2016; Solomon 2016; van Dooren 2014; Weaver 2013).

In the late twentieth and early twenty-first centuries, animals have become safe and apolitical subjects in the way that weather and road conditions used to function in polite conversation in the nineteenth and twentieth centuries. Today, cute cat videos supplant dismal news on smartphones, and National Geographic TV shows have long filled the repressive airwaves of Malaysian state television (Chua 2017; Ngai 2012).

But the animals in this multispecies ethnography are neither polite nor apolitical. They urinate, defecate, and earn their food like the workers caring for them. For the orangutans I describe, extinction threatens the existence of their species, and their extinction will not be the result of their own failures, actions, or inactions.

The emergence of multispecies ethnography means that multispecies ethnographers often stop at the point of wonder that cross-species relations generate in the space of difference, diversity, and multiplicities (Alger and Alger 1999; Candea 2010; Kirksey and Helmreich 2010). Often, such ethnographies incorporate the vantage point of scientists or conservationists, and in so doing privilege the perspective of those with the means to embrace an environmental cosmopolitanism by traveling the world (Braverman 2015; Lorimer 2015; Van Dooren 2014). My perspective approaches multispecies ethnography from a different angle, through its emphasis on geographic specificity and what that entails by way of history, culture, sociality, and ecology, even when that specificity might lead to an emphasis of one species over others. Happy, saccharine stories of multispecies friendship and flourishing are inadequate (Ahuja 2016; Fiskesjö 2017; Freccero 2011; Porter 2013; Tuan 1979). Nor can we be satisfied with stories of human–animal conflict and interspecies competition when our very lives are made possible through symbiosis and coevolution (Haraway 2014, 2016; Mansfield et al. 2014; Subramaniam 2014; Tsing 2005, 2015). We need richer stories that suit the complexity of our times and of our lives.[36]

Ponder, for instance, encounters with semi-wild orangutans. Such encounters are always uncertain. They can lead to an embrace, or to a bloody bite and subsequent infection. They can lead to feeling saliva upon one's body when at the receiving end of an orangutan's raspberry. They can involve the physical impact of a shower of tree branches, the fall of dead leaves like confetti, or to mere avoidance and disinterest. In such encounters, we lose the ability to derive meaning from referential speech. We sense an uncertainty that unfolds from feelings and visual cues generated in the space between bodies—all kinds of bodies, whether animate or inanimate, lively or otherwise. It is here in this relation between humans and semi-wild orangutans that a distinction between affect and emotion makes sense: the shaky sensations wavering between potential joy and potential worry are not merely embodied emotions, for this is where we feel the affective rush of sensation that stirs not from within the body, but between bodies, in the moments before they become emotion, if they become anything at all.[37] This happens on the timescale of seconds and microseconds. It happens with any and all encounters between bodies, but especially in the space of orangutan rehabilitation.

Take, for instance, encounters with the adult female orangutan Ching, who had a reputation for biting local women. Any encounter with Ching was unpredictable. Perhaps the reasons lay deep in Ching's only partially known life history (Braitman 2014). Ching was surrendered to the state's Forestry Department more than a decade ago as a young orphan; she had for years served as an attraction at a luxury hotel in Batang Ai overlooking the man-made lake that had submerged her forest habitat. She had a reputation for disliking women, especially local women, enough to hurt them. Some alleged this stemmed from when a Chinese woman visited the park and refused to give Ching her backpack containing sweets. Others thought it was connected to an incident in 2004 when an intern from a local university teased the captive Ching after her first infant born on site was taken from her. The intern supposedly held the baby and showed her to Ching while Ching was behind bars.[38]

As a Filipina American, I was technically foreign, but often mistaken to be local by Sarawakians. To Ching, my appearance made me vulnerable to attack. This particular orangutan had the power to make me feel reduced to an essentialized subject-positioning, cuing culturally informed ideas of gender and race.[39] The junior officers Cindy and Lin were also vulnerable to her attack.[40] We each responded to that added risk and responsibility differently, as I show in chapter 1. Orangutan bites often require hospitalization. I conducted myself with trepidation whenever I was in her presence.

The International Union for Conservation of Nature issues Best Practices Guidelines for the Re-Introduction of Great Apes as an "ideal code of conduct," unencumbered by the limitations of "location, resources, and government regulations" (Beck et al. 2007: 3). The guidelines define rehabilitation as a temporary condition: "the process by which captive great apes are treated for medical and physical disabilities until they regain health, are helped to acquire natural social and ecological skills, and are weaned from human contact and dependence, such that they can survive independently (or with greater independence) in the wild" (Beck et al. 2007: 5). This definition of rehabilitation is aspirational, just like the guidelines in which it appears: orangutans here cannot be "weaned from human contact." Lisbet is in the second generation of orangutans at Lundu Wildlife Center, while three generations live at Batu Wildlife Center.

The technical term in the primatological literature for Lisbet and her kind is *rehabilitant orangutan*, but they are more accurately described by Layang, Nadim, and their coworkers as *semi-wild*.[41] Semi-wild is a more honest term when release to the wild is uncertain, when sanctuaries are as permanent as poured concrete. Reflecting this reality since privatization, Batu and Lundu are no longer officially named "orangutan rehabilitation centers," but "wildlife centers."

Working at the wildlife center demands an ability to read orangutans not only by discerning individual faces, but by becoming sensitive to such subtler signs as raised hair or how they move their lips. Miscommunication with apes or nearly any other resident of Lundu Wildlife Center could easily lead to painful and bloody bites. Every worker tasked with the day-to-day work of feeding semi-wild orangutans swore that bites are inevitable. Like Ching, some orangutans were repeat instigators of such physical contact.

The potential threat of injury characterizes the work of caring for semi--wild orangutans because there are no physical barriers between apes and people. This contrasts with modern zoos, where experiences with animals take place in a controlled environment and are mediated through the hindrances of iron bars, Plexiglas, or man-made moats. Such barriers define the experience of contemporary zoos throughout the world, whether in Singapore, San Diego, or Sydney. The wildlife center that hosts semi-wild orangutans is different from any other site in the world, as it allows for the layperson's direct and embodied experience of what it is to be at the interface of species loss and vulnerability.[42] In a material way, the orangutan rehabilitation center teaches us how to share a future together amid mass annihilation.[43]

Orang Hutan

Consider the word *orangutan*. Orangutan is often translated as "Man of the Forest," based on the Malay terms *orang* (person) and *hutan* (forest). It comes to English by way of the Dutch physician Jacobus Bontius, who was employed by the Dutch East Indies Company. He offered the first European account of orangutans in the 1600s. On Java, in present-day Indonesia, far from the forests in which orangutans live, Bontius had heard that orangutans are capable of speech but refuse to speak in order to avoid being put to work (Bontius [1642] 1931; Cribb et al. 2014). This has haunting significance insofar that this was in Batavia, when it was an entrepôl of the Dutch slave trade (Baay 2015; van Rossum 2015).[44]

One better versed in Southeast Asia history or philology will tell you that the word *orangutan* was not the colloquial term for people living in orangutan habitats. The famed explorer Alfred Russel Wallace ([1869] 1890) reported that *maias* was the preferred term in Sarawak in the 1850s. It sounded like *mawas*, the common term used on the northern area of the island of Sumatra, across the Karimata Strait (Payne and Prudente 2008). These terms are likely cognates. Contemporary conservationists Junaidi Payne and J. Cede Prudente (2008) note that Sarawakians historically made three distinctions among orangutans: *maias kesa* (or small orangutans or juveniles), *maias rambai* (or medium-sized orangutans, presumably females and subadult males), and *maias timbau* or *maias papan* (or large and flanged adult males).[45] As late as the 1950s, the curator of Sarawak Museum Tom

Harrisson organized the Maias Commission to consider the ape's conservation status. Since then, the term *maias* has faded out of everyday speech in Sarawak and orangutan has taken its place. People working at either of Sarawak's two orangutan rehabilitation centers were unfamiliar with the word *maias*. The translation of orangutan as "person of the forest" or "man of the forest" bears only recent significance for Malay speakers at best.

If we take the history of language into account, we see that the idea of shared humanity through the audial affinity between orangutan and orang likely has its origin in a misunderstanding between a sixteenth-century Dutchman and the Javanese traders with whom he spoke. Understanding the relations that I describe in this book does not require a perspective that centers humanity, such as a Dutch Calvinist vision of animals as degenerate immoral products of the sinful and "detestable" desire of "women of the Indies."[46] Indeed, evolution is irrelevant for how Sarawakian people relate to Sarawakian wildlife. Even without a claim to a shared "family of man"—11 million years have passed since humans shared a common ancestor with orangutans—Sarawakian people and orangutans already share experiences of displacement and arrested autonomy.[47]

Decolonization

Decolonization, as an idea, aspiration, or set of actions, requires a double vision. On the one hand, it requires focusing on the specific contingencies of history, place, and politics. On the other hand, it calls for a comparative view with other forms of decolonization.[48] How might ongoing and future decolonization matter for orangutans in a place where decolonization is usually discussed in the past tense and where independence happened for less than two months, in 1963? This question requires thinking about specific space and the people and politics that have helped shape this place.

Sarawak has hosted humans for millennia.[49] Written records indicate that from the 1300s CE, Sarawak was at the periphery of maritime empires, first the Java-centered Majapahit Empire and then the Sultanate of Brunei two centuries later (Blussé and Gaastra 1998; Nagata 2011; Reid 2000). Coastal vassal settlements paid tribute to these empires, while upland people in the interior of Borneo participated in extensive trade networks that appeared to reach beyond Borneo, extending into China, perhaps as far back as 800 CE.[50] Sarawakians with Chinese heritage can trace their ancestry in Sarawak for more than two centuries. In the 1600s CE, Ibans from the interior of Borneo expanded their sovereignty by waging war in present-day Sarawak (Dimbab et al. 2000; Jawan 1994).[51]

The era of European colonialism began in Sarawak in 1841, around twenty years after Sir Thomas Raffles founded Singapore as a trading post serving the British East Indies Company.[52] James Brooke, the son of a colonial judge in British India, was inspired by Raffles to found a port to serve British maritime trading interests. In turn, the famous imperialist author Rudyard Kipling found a muse in James Brooke and his temerity. Kipling coined the verb "Sar-a-whack" to describe how an Englishman became the divine king of a land on the outskirts of the British Empire (Kipling 1919).

The historical record supports no such story of deification, but instead points to a story of subterfuge and gunboat diplomacy.[53] In 1841, on behalf of the sultan of Brunei, Brooke suppressed a rebellion in the coastal city of Kuching, at the mouth of the Sarawak River. He then demanded that the sultan cede to him the area's control. Brooke's power was concentrated in the area of the city of Kuching, and his rule over the kingdom of Sarawak was solidified by the use of excessive force, for which he faced charges in Singapore, on which he was ultimately acquitted (Brooke and Drummond 1853; Hume 1853). One of his policies, continued by his heirs, was to suppress headhunting in general, but also to encourage the practice when it suited the Rajah's expansionist agenda (Pringle 1970). By harnessing such rituals, Brooke attempted to arrest the autonomy of headhunting people.

For a century, Brooke and his heirs autocratically ruled Sarawak. They fashioned themselves in the model of the British Raj by giving themselves the title of rajah.[54] They were known by the racially marked appellation "The White Rajahs," which distinguished them from rajahs of the Indian subcontinent, who were subject to indirect rule by Britain. The White Rajahs were more autonomous. James Brooke's nephew Charles Brooke (rajah, 1868–1917) inherited the throne, expanded the territory of the Raj, and made it a protectorate of the British Empire as the global timber industry boomed. One manifestation of this boom was the eventual development of Batu Nature Reserve, on which Batu Wildlife Center is located.

On the centennial of his ancestor's control of Sarawak, Charles Vyner Brooke (rajah, 1917–46) promised eventual sovereignty and independence for Sarawakians.[55] He did not fulfill this promise. Three months later, the Imperial Japanese Army invaded Sarawak. The industries that made

Charles Vyner Brooke wealthy, namely oil and rubber, were of supreme interest to Japan in its effort to colonize Asia. Under these tumultuous circumstances, Sarawak's independence was never realized.

After the Imperial Japanese Army occupied Sarawak from 1941 until 1945, and after a period of British martial law, Charles Vyner Brooke returned from exile in Australia in 1946. Instead of officially abdicating power and granting Sarawak its independence as he had promised before the war, he ceded Sarawak to the British Crown and personally gained a third of Sarawak's financial reserve. Through this exchange, Sarawak's independence was further stymied. It remained a British crown colony from 1945 to 1963, even as decolonization officially ended elsewhere in the British Empire (Porritt 1994).

Contingency seems to have brought Malaya, Sarawak, Sabah, and Singapore together in 1963 when they became the federal state of Malaysia, although Singapore left the federation a few years later to become an independent city-state.[56] The two land bodies of Borneo and the Malayan peninsula are separated from each other by hundreds of nautical miles. Until 1981, the two places had two different time zones. Sarawak is now a semiautonomous state of Malaysia. Its semiautonomy is expressed through its own immigration policies (He et al. 2007).

The distinction between Malaya and Sarawak is not just spatial. Sarawakians proudly lack a history of ethnic violence in lived human memory, unlike peninsular Malaysia or on the other side of the land border shared with Indonesia (*Kalimantan Barat*)—although anticommunist actions in twentieth-century Sarawak racially targeted Hakka Chinese minorities (Kua Kia 2008; Peluso and Watts 2001; Yong 2013). Sarawak is home to around thirty ethnic groups, the largest of which is the Iban. To be a local in Sarawak is to be at ease with multiculturalism, religious plurality, and other forms of difference.

Sarawak's relationship to the federal government is more fraught; although the territory is rich in natural resources, its human population is poorer than the peninsular Malaysian states. A point of contention, even for Sarawak's established politicians loyal to Malaysia's dominant political coalition, is that Sarawak retains only 5 percent of revenues from the oil and gas it produces (Yong 1998). Even after the fiftieth anniversary of Malaysia, the official holiday of Malaysian independence continues to be celebrated on the day that only the peninsular states of Malaya—not Malaysia—

became officially decolonized.[57] Disgruntled Sarawakians of all ethnicities continue to debate the merits of gaining independence from Malaysia, yet recent electoral politics show acquiescence to the status quo.[58]

Most visitors who come to either of Sarawak's two wildlife centers to see semi-wild orangutans, including volunteers who come to perform hard labor for Sarawak's orangutan rehabilitation efforts, are unaware of this history. Yet the details matter greatly for the orangutans under their gaze and in their presence. It matters that indigenous people were evicted from the forests in which these orangutans now live, in the 1860s during Charles Brooke's reign as the second white rajah and then in the 1980s in postcolonial Sarawak, now a part of Malaysia. It matters that the trees these orangutans climb were planted in the 1920s for scientific forestry to extract more wealth from Sarawak and at the expense of original orangutan habitats. It matters that the monocultural forests along the highway route to Lundu are actually feral rubber trees that were planted by forced labor during Japanese military occupation. And it matters that the little space these semi-wild and captive orangutans now have is carved away by other interests, like the construction of a larger airport using sand mined at the edge of Batu Wildlife Center. Thus, Lisbet's failure to demonstrate independence within the confines of the park, or wildlife ranger Nadim's talk about the material constraints on the orangutan Wani's autonomy, as I describe in chapter 5, are part of a larger story about Sarawak's arrested autonomy.

Arrested autonomy is expressed in Sarawak's semiautonomous status. It is conveyed by orangutans who seem to be able to roam freely, but are actually constrained in a space shaped by colonial interventions on the land. It is evoked in Layang's conviction that one ought to do something but is instead actively prevented from doing so. Arrested autonomy is arrested decolonization in the face of ongoing colonialism when colonialism is supposed to be over. It is the frustration of having the means intended to foster independence instead work toward continued dependence. Such forms of arrested autonomy serve as a recurring trope in Sarawak's history since colonial contact.

The feeling of arrested autonomy is perhaps familiar to some readers, though it surfaces in different ways, with different figures, in different circumstances. Recovering drug users released after rehabilitation experience arrested autonomy (O'Neill 2013). Anyone who has been stuck in an institution has felt its limits pressing down even while being told that contraction is in preparation for a future expansion. The constraints on their freedom that orangutans experience and the constraints on their caretakers are related and to some extent shared.

Readers may feel uncomfortable with the idea that arrested autonomy could be shared between orangutans and people, especially when people who are racialized as native or who are denied the dignity of human rights are often treated "like animals." But the insult of animality and the deprivation of humanity both depend on a colonial hierarchy in which some people are treated as less human than others (Weheliye 2014).[59] Rejecting colonialism also requires rejecting the refusal to acknowledge the possibility of shared experiences with nonhuman others, for lack of a better word. In other words, decolonization offers potential recognition that colonialism has brutal impacts for many of Earth's inhabitants, many of whom are not human.

Rehabilitation

Lundu Wildlife Center is an orangutan rehabilitation center. Yet it looks remarkably like a zoo. Reconciling its appearance with its practices takes work, as I found out in a 2010 conversation with a commercial volunteer.

I stood on the viewing platform above the orangutan enclosures. A blond woman in her twenties toting a camera joined me as I finished jotting notes. She was an ecotourist finishing her monthlong volunteering stint, which cost about US$4,000, excluding airfare. She struck up a conversation with me. I posed a question to her:

JUNO: How do you make sense of all this captivity?

VOLUNTEER: I came in with my Western hat on, having seen only babies and mothers, never having seen these huge males. With my Western hat, you think forest all around, why can't we release them? But then you get here, you have to put on a local hat and see how it's so complex, that you can't just release them. *It's sad to say this, but it's like rehabilitating sex offenders.* You can't just release them back into society. They need rehabilitation. It's just so complex. [Pause] Everything's a catch-22! They're wild animals that can injure you. When I show these photos to my friends they're like, "Why all the cages?" And I have to explain that before releasing them, there needs to be some level of captivity and then you move on to semi-wild and then hopefully to a wild state. [My emphasis]

By equating displaced orangutans with sex offenders needing rehabilitation, this volunteer tried to reconcile the difficult truth that the orangutans she came to know actually spent most of their time in various states of captivity.[60] Her equation enabled her to imagine confinement in cages as a prison and enclosures as temporary; the orangutans' freedom would be gained gradually, until they were finally able to earn free-range autonomy in the forest.

Our conversation occurred years before international outrage at the news of a brutal and ultimately fatal gang rape on a public bus in India in 2013. At the time, neither of us could imagine how the sexual behavior of orangutans would be highlighted in international news media, when they were used to suggest that some individuals were biologically inclined to rape others—by extension, an explanation for human brutality and misogynistic cruelty (Lenin 2013). Indeed, biologists like Wrangham and Peterson (1996) reference orangutans as an example of a species in which forced copulation is a reproductive strategy "found in nature."

By likening the adult male orangutan under our gaze to a sex criminal, the volunteer made explicit the way orangutans' sexuality could be used to justify their captivity. Seeing male orangutans as sex offenders naturalized the violence female orangutans experience when human activity physically confined them in the wildlife center. Such naturalization drew attention away from the human interventions that made this a space of "nature-culture" in which "nature" was performed in a built environment, built through a long history of human–animal–plant encounters (Cronon 1996; Haraway 2003; Sivaramakrishnan 1999; White 1995).

The volunteer's sentiment also expressed the idea of prison as a form of correction instead of punishment. This view perceives crime as an exercise of free will and moral turpitude requiring rehabilitation and not the result of an unfair justice system (Alexander 2010; Baldry et al. 2011; Davis 2003, 2012; Gilmore 2007; Kornhauser and Laster 2014; Tonry 2001). The rehabilitation center for orangutans, then, at least in the mind of volunteers like this one, was a promise of eventual release upon good behavior or penance in a species' penitentiary. Yet what was being corrected was also seen as inherent (Pandian 2008). As she said, semi-wild orangutans were still "wild animals that can injure you." How could rehabilitation work if it was fighting something innate?

The volunteer's ability to reconcile with the point that orangutan rehabilitation was like rehabilitating sex offenders made me feel that orangutan rehabilitation was a deeply colonial project in a postcolonial place, even as the practices of orangutan rehabilitation evoked decolonizing possibility. The dream of achieving an eventual goal of freedom meant for her that the day-to-day life of captivity and constraint was perfectly acceptable. She took comfort in arrested autonomy.

Her desire for the orangutans' physical freedom was ultimately constrained by her strong demand for personal safety. That vision of orangutan freedom would push orangutans to enter "society" when orangutans as a species would likely reject such social belonging. Postcolonial governance in the form of land administration and tourist visas limit her colonial aspirations for achieving the eventual independence of others: no, they cannot just be released when forest is all around and when she thinks it's fit. No, she cannot stay indefinitely in Sarawak to see if or how such a release to society occurs. If she were to reject the wish for safety, if she abandoned the metaphor of rehabilitation as a penitentiary, if the thrill of temporarily visiting the interface of extinction became the dull pain of empathetically sensing another's suffering under conditions that look like her definition of happiness, what could be otherwise?

Iban workers had different attitudes than this volunteer. Kak had been employed by the volunteer company to work as the orangutan nanny. She lived in the Iban longhouse where Apai Julai was the headman, and she hailed from Ulu Sebuyau, the area where Alfred Russel Wallace was based during his explorations of Sarawak a century and half before. Kak explained that orangutans evoke the feeling of *geli*, a creepy abhorrence, the feeling of seeing something close up that you should not be seeing. Layang, the unofficial expert on site, recognized the orangutans' distinct personalities—Ching could whistle and she once hid a key in her mouth for two days; James was a huge male with flanges beginning to develop on his cheeks, yet he was afraid of groups of people. If semi-wild orangutans were prisoners, they were wrongly confined. For Layang, and for the people he influenced with his ideas and practices, rehabilitation was not about suppressing orangutan behavior, but about experimenting with ways of inhabiting the same space—not by imposing a sense of safety, as the volunteer supported, but by embracing the risk of vulnerability. Inspired by Layang, I feel that decolonizing extinction would require letting go of the aspiration for a safe inequality, even if one risks experiencing pain.

Chapters

This book requires its readers to recognize place, time, and circumstance: How are people, animals, plants, bacteria, and other earthly bodies encountering one another at a given moment? What possibilities are generated when they are together in the same place and time in a fleeting moment, whether that shared moment is over seconds or millions of years?[61]

The embodied relations that happen at these wildlife centers push us to consider how to live and die with others. We can no longer entertain the fantasy of autonomous, isolated living—as seductive as that fantasy may be when we want to picture orangutans roaming freely in the grand forest canopy. Instead, the interface between displaced orangutans and the people caring for them teaches us that living together, when our existence is threatened by slow but cataclysmic transformations, entails becoming vulnerable to one another, risking even the possibility of losing our own lives.

The giving in to risk and sacrifice occurs in a context of violence, where some who exist on this planet, including some who are also human, are more readily subject to force, manipulation, and imposition than others. These forms of violence are expressed through variations of intersubjective and structural relations: orangutan habitat loss resulting from the colonial and later the postcolonial extractive economy of forestry and sand mining, the forces that pushed indigenous Sarawakians to survive through wage labor and to indefinitely defer independence for all Sarawakians, or global capitalist inequalities that volunteers try to ease and anthropologists like me try to understand.

My goal in decolonizing extinction is not to transfer the power of decision making to newly appointed experts who better understand such violence. Replacing a timber tycoon with a conservationist at the top of the chain is not enough. Rather, I believe that decolonizing extinction requires a fundamental reorientation toward others, especially nonhuman others, in which we accept the risk of living together, even when others' lives pose dangers to our own. I make these arguments in six chapters divided into three sections.

Chapters 1 and 2 together examine how people build social relations with members of a species famous for a love of solitude. Such sociality is embroiled with differences forged by colonial hierarchies, political economy, evolutionary distance and notions of race, gender, sexuality, and species. Relations are generated through a contingently shared interface.

Chapter 1 examines the first-ever orangutan rehabilitation experiment run by Barbara Harrisson of the Sarawak Museum. I interpret her "ape motherhood" as an effort at instilling independence among orphaned orangutans in the 1950s and 1960s, in the midst of debate around Sarawak's independence following official decolonization. This chapter traces how the ideology of "ape motherhood" was replaced by the contemporary concept of "tough love" and shows how both ideals are informed by ideas of gender as well as colonial and postcolonial conditions of labor.

Chapter 2 considers how affect, sensed on the surface of skin and grounded in a specific planetary surface, generates a global economy of commercial volunteerism. Examples of affective encounters include the everyday, ordinary, and yet extraordinary chores of the wildlife center staff and volunteers: evacuating orangutans, cleaning their cages, and carrying out hard manual labor. Even the technological mediation of "crittercams" cannot replace the experience of bodily presence with a member of an elusive and endangered species in the same space and time.

Chapters 3 and 4 together consider the problem of enclosures as experienced by both wildlife and their caretakers. "Enclosure" refers to the place where captive wildlife dwell. It is synonymous with "exhibit," although the concept of enclosure is more oriented toward the animal being housed, while the concept of exhibit is more oriented toward its pedagogical function for human visitors. The concept of enclosure also implies the dismantling of commonly held lands; the subsequent displacement, dispossession, and eviction of peasants; and the push for them to become wage laborers as a means of survival (Grandia 2012; Marx 1981; Polanyi 2001; Thompson 1968, 1975). Enclosure is shorthand for a shared interface of loss between displaced wildlife who have nowhere else to go and displaced people with few options but to work for wages to survive.[62]

Chapter 3 considers the impact of space on the social lives of orangutans. Semi-wild orangutans live in insufficient space, which, paired with the human tendency to blame extinction on insufficient sexual reproduction, exacerbates the problem of forced copulation that female orangutans are made to bear. The ranger Nadim succinctly describes the semi-wild orangutan's state as "free but fearful." How people on the ground justify a system of sexual violence shows both the possibilities and the limits of empathy. It compels us to consider how we use the word *rape*.

Chapter 4 examines the transformative loss shared between displaced indigenous caretakers at Lundu Wildlife Center and the animals under their care: crocodiles, turtles, sun bears, as well as orangutans. Both wildlife and their caretakers must "*cari makan*," which translates from Malay to English as "find food" and is used as a local idiom for wage labor.[63] In this chapter, I am interested in the ways wage labor forces the alienation of people from animals, replacing older notions of them as omens and kin with new kinds of knowledge.

Together, chapters 5 and 6 consider a range of possibilities for a future in such constrained conditions. Futures here have a double meaning: both financial capitalization when orangutans are considered assets from which to draw future profit through their scarcity and the liberatory futurism of decolonization. I conclude the section by engaging the dilemma of what kind of future is possible when we live with nonhuman others whose livelihoods are simply deadly, like endemic microbes that are also bioterrorist agents.

Chapter 5 returns to bebas, the word that caretakers use to describe semi--wild orangutans' freedom. Bebas, which I translate as physical autonomy or freedom, offers a theory of decolonization, one that shows how autonomy is currently arrested for Sarawak's orangutans and for Sarawak's people. This chapter considers the scope of the longue durée by pursuing the etymology of the terms that we use to envision political futures.

Chapter 6 unpacks an idea inspired by a volunteer's grandmother: the wildlife center operates as a hospice for a dying species. Hospice is a useful analogy, considering it as a place of care when freedom outside of confinement ceases to be possible, in which caregiving is compensated with wages, in which both caregivers and care receivers are vulnerable to harm, and in its operations as a commercial or for-profit institution. The analogy reaches its limit at the point when caring interventions stop and death cannot be willed away.

Loss evokes the pain of absence—that flash of inconsolable longing when you contemplate objects that have outlasted loved ones (Rosaldo 2013).[64]

Annihilation is loss amplified on a scale so vast that it is hardly fathomable (Masco 2006). Unlike nuclear holocaust, the extinction of orangutans poses no existential threat to all of humanity. Yet the fact that orangutans' survival in the future is contingent on human actions illustrates all too well that our own existence on this planet is shaped by relations with others, including and especially nonhumans.

This is not a time to fatalistically give up on caring how others try to eke out a living under dire circumstances. My purpose here is to encourage alternatives to what projected futures might hold by highlighting what can be observed. My hope is that we can seriously consider what could be otherwise.

PART I **RELATIONS**

Chapter 1

FROM APE MOTHERHOOD TO TOUGH LOVE

When explaining the project of orangutan rehabilitation to a potential donor, the British program manager of Endangered Great Ape Getaways (ENGAGE) Tom said, "Our philosophy is to provide tough love, not mother's love."[1] He went on to mimic what he would say to an orphaned orangutan sent to the center: "So your mother's been shot. Tough luck; off you go." He stuck out his hand in front of him and deliberately pushed the air, gesturing what such a training of rejection looks like. How did motherhood come to be the antithesis of rehabilitation, especially when it had been the idiom of choice for Barbara Harrisson, whose experiment in raising orphaned orangutans began the practices used in all orangutan rehabilitation efforts today?

Harrisson, as the wife of the curator of the Sarawak Museum of Natural History and Ethnology, raised orphaned orangutan infants from 1956 until 1967. Unable to return them to the wild, where they would die without their mothers, where "the wild" was subjected to mass logging and road construction hitherto unseen, and unwilling to send them to metropolitan zoos built during the late nineteenth century, where orangutans in captivity were housed in concrete exhibits and displayed signs of poor health, Harrisson sought an alternative. She worked to realize that alternative with the help of a small workforce active in her home.[2] This experiment took place at the periphery of modern biology, in the space of the colonial domicile.

Domestic labor was crucial in this domestic laboratory, which was the site of active experimentation looking toward a flourishing future independence in a decolonizing Sarawak. This experiment with orangutans meant their learning to live independently while living together with human caretakers. For Bidai and Dayang, who worked in Harrisson's home, it meant figuring out how to foster independence in others while they experienced new forms of dependency on money through wage labor. For Barbara Harrisson, it meant staking out new kinds of social relations and an attempt to achieve influence from a distance by way of modern scientific knowledge production. For the orangutans and the people who cared for them, this was the direct interface of extinction, a condition new to everyone in it.[3]

Fifty years later, the term *ape motherhood* had been replaced by *tough love*. Both concepts employed the same tactics of rejection, in which handlers literally push away orangutans. Both occurred at the site of orangutan rehabilitation, in the space between human handlers and infant or juvenile orangutans. Both were enacted at the peripheries of scientific knowledge production. Both strove for orangutans' eventual independence. Yet the earlier domain of ape motherhood transgressed the boundaries between the domestic and the scientific, between private home and public museum. Tough love cut through various boundaries of private and public. It occurred in a space created by a private–public entity, shared between a private volunteer company and the semigovernmental, privatized wing of the Forestry Department. It was a space that rejects the home and the domicile in favor of the wildlife center: a privately managed but state-owned space, a place of work, not of dwelling, a place of manly bravado and not womanly motherhood—regardless of how transgressive that motherhood may have been.

The comparison I make across a time span of fifty years is not meant to demonstrate how far we have progressed. Indeed, it shows the opposite: the challenges Barbara Harrisson faced continue in the present: how are humans to conduct themselves in the interface of extinction? What futures are possible in these circumstances? The responses to these questions have evoked the same issues in both the past and present: it compelled, and continues to compel, the production of scientific knowledge at the peripheries of science.

In this chapter, I examine the embodied colonial legacies of caring for orangutans. Barbara Harrisson's idea of ape motherhood was carried out in collaboration with her assistant, Bidai anak Pengulu Nimbun, a Selakau youth who lived with the Harrissons, and to a limited extent Dayang, Harrisson's matronly Malay housekeeper.[4] I show that Harrisson's ape motherhood was both a colonial project of stewardship and one that envisioned an alternative future that stood against the institution of the bourgeois colonial family. I then examine a confrontation described by Harrisson in her memoir *Orang-Utan* as a way to explain how she ambivalently reflected upon care, gender, ethnicity, and expertise in the colonial era. I then end with a comparison between contemporary and past ideas of orangutan infant care, as they were mediated through the language and embodiment of tough love and through the method of photo elicitation, in which contemporary orangutan handlers interpreted two archival photographs from Harrisson's orangutan rehabilitation efforts. I ultimately argue that Harrisson's interspecies experiment to instill freedom among displaced orangutans was both a colonial and a decolonizing intervention, one that served as a history of a feral future. Yet that history is dissociated from the present, even as the present evokes that history through everyday violence.

Motherhood across Species

The project of handling displaced infant orangutans fell into Barbara Harrisson's lap, literally, on Christmas Day in 1956. The Forestry Department had confiscated an illegally kept infant orangutan and sent the little one to Tom Harrisson, who brought him home. Their home, Bunglo Segu, was unlike any other British colonial abode. It was the modestly sized summer bungalow of the last White Rajah of Sarawak before he abdicated the throne following World War II and transferred power to the British Crown. As the newly appointed curator of the Sarawak Museum following the war, Tom Harrisson relocated the entire structure in 1947 from Serian Road to its present-day location on what was known during the colonial era as Pig Lane. Harrisson commissioned elaborate Kelabit and Kenyah murals on the walls and ceilings. Sarawakian artwork and artifacts filled the entire house, blurring the distinction of spheres between private residence and public museum as well as between private possession and public ownership.[5] The house had accommodated Harrisson and his Kelabit wife

Sigang, who as a Kelabit and thereby a Sarawakian native, was unable to live with him in the government resthouse. By late 1955, Sigang and Tom Harrisson's marriage had dissolved; he and the then-named Barbara Brünig nee Güttler developed a close relationship while they worked together at the Sarawak Museum. Barbara divorced her first husband and married Tom Harrisson in May 1956.

With the arrival of that infant orangutan six months after Barbara and Tom Harrisson married, the house was to become the site of the world's first-ever orangutan rehabilitation project. This was twenty years before the formation of Sepilok Orang Utan Reserve in Sabah and Biruté Galdikas's even more famous Tanjung Puting site in Central Kalimantan, Indonesia. Barbara would go on to raise a total of twenty-five orangutans at the home she shared with Tom Harrisson, until her departure from Sarawak in June 1967, when she was forced to leave following official decolonization.[6]

Barbara Harrisson fostered infant orangutans so that they might eventually lead independent lives. This occurred during Sarawak's years of political transition from British crown colony to the federal state of Malaysia, between 1961 and 1966. Her initial musings about orangutan independence occurred a year after the Asian-African Conference took place in neighboring Indonesia in Bandung, where the Third World stance against "colonialism in all of its manifestations" was articulated (Abdulgani 1955). This was fifteen years after the last White Rajah promised Sarawak's eventual self-governance in 1941, ten years after he reneged on that promise and transferred Sarawak to Britain in 1946, and seven years after an anticolonial assassination of the first British appointed governor of Sarawak in 1949. As a form of punishment, the colonial government forbade the participation of Malays in political parties (Leigh 1974). The problem of how to instill independence was as much a political one for Sarawak's colonial society as it was a programmatic one in the Harrisson household.

Barbara Harrisson's experiments on how best to care for orphaned orangutans show how the colonial dynamics of care and intimacy shaped orientations toward native animals that by then were categorized as endangered species (Fisher et al. 1969; Harrisson 1965a, 1965b, 1987). Orangutan infants are not just vulnerable beings, deprived of their mothers who would otherwise have cared for them for about seven years (Galdikas and Wood 1990). Their bodies bear the weight of being archives for a lively future worth saving from potential extinction and worth fostering so that they may eventually return to a life with some form of freedom.

The infant orangutans themselves make their own material demands: satiating their hunger is one of their needs; grasping arms accustomed to holding onto the furry flesh of their mothers' bodies makes a demand, too. Such demands elicit complex responses. Barbara Harrisson described one such response in her memoir *Orang-Utan* as "ape motherhood." This kind of motherhood was a form of interspecific care that fostered relations of rejection instead of affection. Its authority rested on Harrisson's appropriation of scientific instruments and her own fraught relationship to the colonial state as the German-born wife of a British colonial officer in which her whiteness represented the British power to police and punish Sarawakians of all ethnic groups.[7]

Barbara Harrisson became a researcher and conservationist by carrying out the work of research and conservation at a time when formal education had been unnecessary. In fact, her first degree in art history meant that her formal education exceeded her husband's, since he had left Cambridge University in his first term and never resumed his studies. Without an advanced degree, Barbara Harrisson personally excavated Niah Caves, which at the time had been the site of the oldest known hominid remains in Asia. She did this while carrying out what she called an "experiment" in fostering orangutan infants out of her modest home, which had been partially converted into a rehabilitation center and later a section of the newly founded Bako Nature Reserve. When she left Sarawak, she accepted an invitation to pursue a PhD at Cornell University, where she specialized in Asian ceramics. Although she was in correspondence with the newly formed Wenner-Gren Foundation, the possibility of studying nonhuman primates at a Max Planck Institute in West Germany with Wenner-Gren funds did not materialize. She went on to direct the Royal Museum of Ceramics in the Netherlands.[8] There, she retired and continued to live independently in her own home until her death at the age of ninety-three in 2015. When I met her in 2006 at her home in the Netherlands, I was struck by her commanding height, the clarity with which she spoke, her generosity in letting me stay overnight in her home, and the precision of her words: "I will not speak about my personal life with a stranger."

The wish to not discuss her personal life may seem to be an espousal of bourgeois private and public spheres. Yet her own memoir of her experiments in rehabilitating orangutans made no such distinction between the personal and the public. Originally published in 1962, Harrisson's book *Orang-Utan* (1987) takes us to the most intimate spaces of her life in Sarawak: the bed on which she lay sick with a cold when Bob, the first orangutan she raised, was brought to her lap; the bathroom in which she and her husband brushed their teeth and hair while infant orangutans would stare at them from the other side of the wire screen covered window, where the infants slept on the veranda in cages with burlap gunnysacks. Her writing evokes the memory of Alfred Russel Wallace and the genre of the travelogue in which natural history and ethnography blend together. Yet her descriptions are set amid sweeping changes brought about by modernization. Her vision of Sarawak is marked with a careful respect for the details of her everyday encounters with Sarawakians, from the city's Chinese coffee shops, to its rural Iban longhouses, to the grounds of her own home.

Her refusal to address my personal questions was likely less about the preservation of public and private distinctions, and more about a refusal to respond to the stories in Tom Harrisson's biography, *The Most Offending Soul Alive*, written by a member of their colonial social circle in Kuching (Heimann 1998). The work details Barbara Harrisson's life and the life she shared with Tom Harrisson. My interest here is not to pry into her marriage and intimate life for the sake of unearthing scandal. Instead, my interest is in how colonial and transgressive intimacy shaped her own and her assistants' conduct in raising orphaned orangutans, which was to become the standard for other orangutan rehabilitation sites. Their conduct served as the basis of how to live together at the interface of extinction.

In her memoir *Orang-Utan*, Barbara Harrisson distinguishes her household and by extension herself from the hegemony of the colonial bourgeoisie. This is especially expressed in regard to the aesthetics of the home. She explains that her home was "the housewife's bad dream, a conglomeration of all things Bornean, pinned and stuck on wall and ceiling, lying on tables and floor everywhere and nowhere" (1987: 32). Her tastes stood in contrast to "lady visitors of orthodox tastes," who would enter her home and either be shocked into silence or compelled to pose the stark question, "How can you *live* here?" (1987: 32). Through her portrayal of such visitors, Harrisson rejects the civilizing authority of British colonial femininity (Burton 1994; McClintock 1995).

If the interior spaces of their home rejected Western civilization and celebrated Sarawakiana instead, the same sentiment echoed through the "green wilderness" of their bungalow's exterior (1987: 32). It was not severely manicured European landscaping; neither was it the controlled flourishing of an English garden (Thomas 1996; Tuan 1984). Rather, it was as Sarawakian as the rest of the house: durian trees and pineapple bushes with frangipani and hibiscus. It was technically not a "wilderness" since her descriptions consisted of cultivars, domesticated by definition. Yet at Bunglo Segu they took on a life of their own. Harrisson illustrates feral growth that was made possible by the "sporadic" quality of human interventions on the land (1987: 32). This too stood in distinct contrast to the gardens of Borneo's swidden agriculturalists, including Iban, Bidayuh, Kelabit, and Selakau peoples, whose carefully managed gardens featuring rattan and durian trees, coming in and out of years of production and lying fallow during rest periods, are often misunderstood by outsiders as abandoned or lying waste (Fried 2000; Merchant 1980; Padoch and Peluso 1996; Scott 1998; Tsing 2005). Harrisson's colonial home and garden were neither civilized nor wild, but something else: something odd and extraordinary in its alterity against colonial family and propriety.[9] Her garden mirrored the feral futures for which she hoped on the behalf of the orangutans she raised.

Adding to that "something else" was Barbara Harrisson's lack of children, which distinguished her from her female peers among the colonial elite in Kuching. It was also a major point of difference between herself and the Sarawakian woman with whom she had the most contact: her housekeeper or *amah*. Harrisson perceived that her Malay housekeeper, Dayang, wanted them, her employers, to have a child. Harrisson writes, "In her [Dayang's] view, a house without children was no house, empty, without purpose, without fun" (Harrisson 1987: 38). Following Ann Stoler and Karen Strassler (2000), it is quite possible that Dayang's interest in children could have been an interest in the security of her employment within a colonial household. To the contrary, Harrisson seems convinced that Dayang's interest was an emotional or personal one tied to reproduction as a form of female fulfillment.

Readers see what Harrisson perceives as her housekeeper's genuine interest in children and reproduction upon the arrival of Eve, the second orangutan at their house. Eve was extremely malnourished at just seven pounds, despite having teeth that indicated an age of about ten months. Perhaps it was the sight of Eve's skinny body or the chafe marks on Eve's neck from a chain that left the skin raw that compelled Dayang to say something. It is in this moment that Barbara Harrisson explains to readers that her housekeeper has raised many children, alluding to the idea that Dayang raised her own kin and not her employers'. Harrisson writes that Dayang reared "a crowd of children and adopted children" (1987: 38). The multitude indicated by the word *crowd* could have been meant to convey Harrisson's own alienation toward child-rearing. It could also have been meant disdainfully to convey masses of individuals, like a mob, or the litters birthed by such domesticated and commensal animals as cats, dogs, and mice. Yet for Harrisson, it works to establish a difference between an affectionate and tender kind of motherhood versus a form of care that requires being tough and stern. Upon seeing the sickly and skinny Eve, Dayang in Harrisson's memoir, "looked her over with suspicion: 'She is surely small, Mem, she must take milk or she will die—let me try and take her.' With a sigh of relief I [Barbara Harrisson] handed her over, hoping that Dayang at last had found what she missed so much in our household: a baby to cuddle and look after" (1987: 38).[10] In that moment, Dayang represents a sympathetic human motherhood, in which feeding and cuddling are crucial for the survival and well-being of the desperately dependent baby. This is set as the norm against which Harrisson finds her own way.

The expression of care performed by Barbara Harrisson, which she implicitly contrasts with Dayang's, echoes the science of motherhood popularized in guidebooks written by Western medical practitioners in the first half of the twentieth century. Child-rearing guidebooks shifted the authority of mothering from networks of women to trained medical experts: from social hygienists Keller et al.'s *Kinderpflege-Lehrbuch* (1911), which was part of the German and Anglophonic eugenics movement seeking to professionalize the care of children, to Dr. Spock's influential international bestseller *Baby and Infant Care* (1946), first translated into German as *Your Baby, Your Happiness* in 1952 (Spock and Kaspar 1952), which reflected the increased influence of medical specialization in rearing children (Weiss 1977).[11] While the shift in expertise of motherhood happens over decades in the American, British, and European contexts, these different paradigms literally face off in the embodiments of care offered by Dayang and Harrisson.

Readers of the book are shown the failure of Dayang's kind of motherly tenderness. Harrisson writes: "Eve refused to drink and Dayang registered despair: 'She will die, Mem, you will see!'" (1987: 38). Eve's refusal prompted Barbara Harrisson to follow a new course of action and pursue a new way of becoming. Harrisson writes, "Next morning, I made up my mind. With a glass pipette I forced milk into her. After some spitting and coughing, violently trying to get away, she took it; slowly first, hesitatingly. Then I changed over to the baby's bottle and she really drank. The first battle was won" (1987: 38). Described through a militaristic metaphor, a scientific instrument became a tool of domination used for the sake of the survival of the dominated. Force-feeding is an act of violence intended to perpetuate life and end protest, often at the expense of the agency of the one refusing to eat (Khalili 2013; Reyes 2007).

Harrisson's form of motherhood exceeds the expectations of scientifically influenced infant rearing. The glass pipette was an instrument of the modern scientific laboratory, made to administer a precise volume of liquid. Using it to feed Eve in 1957, Harrisson transforms her home into a laboratory. Their home, as feral as it appeared, became an active site of scientific knowledge production—not one merely subjected to outside medical expertise.[12] In the course of the memoir, Harrisson, as the experimenter and scientist, used many tools on Eve and other infant orangutans. Their teeth and the eruption of new dentition became objects to measure and count. The bathroom scale was transformed into an instrument that facilitated the charting out of their changing weights. The chart became a way to view and quantify the metrics of their bodies.[13] The pipette, dentition chart, and other measurements that were used to view the infant orangutans anew transformed the care of infant orangutans from what could have been a project of universal and trans-specific motherhood, regardless of human or ape specificity, into an experiment in scientific inquiry: How does one sustain infant orangutan life without an orangutan mother? The question relied on the instruments of modern science; the answer relied on the domination of Eve and the other infant orangutans receiving care.

Eve refused to drink at first. She was also reported by Harrisson as violently trying to escape being fed.[14] Dayang interprets Eve's rejection of food as a rejection of life. Dayang accepts this, while Barbara Harrisson does not. Harrisson prevents that rejection by force-feeding Eve. In this respect, force-feeding Eve is a forced preservation of life. The possibilities of Eve's survival are constrained by the parameters set by her relationship with Barbara Harrisson and Bidai, who become authorized to administer her care and survival. Dayang's sympathetic, human motherhood is at its limit when she accedes to Eve's refusal to eat. She seems to construe Eve's refusal to eat as a refusal to survive under present conditions.

Survival for Eve means attachment to Barbara Harrisson's intentions. In that moment, Eve's encounter with Harrisson and her glass pipette constitutes Harrisson's new ontology as a scientific authority on orangutan infancy. Likewise, Harrisson's encounter with Eve constitutes Eve's ontology as an orangutan infant reared by humans, whose conditions of living henceforth will be shaped by her encounters with Bidai and Harrisson.

The relationship between Eve and Harrisson is mutually transformative and appears to be an example of what Vinciane Despret (2004) calls "anthropo-zoo-genic practice." This phenomenon names the "new manner of becoming together, which provides new identities" (2004: 122). Anthropo-zoogenesis is contingent on the idea of availability, that is, of each party being available "to affect and be affected" by the other, which she distinguishes from docility, in which submission is forced. Docility for Despret occurs in relations of care when the caregiver imposes changes in others without herself being changed by the encounter. Yet even though Harrisson and Eve are mutually transformed in their relationship to one another, Harrisson nevertheless forces Eve to drink, and with that force she gains Eve's docility. Here I must differ from Despret: the relations between Eve and Barbara Harrisson force me to see that Despret's distinction between docility and availability is a false abstraction. The story of Eve with Harrisson shows us that relations of domination and discipline are mutually transformative among those available to being changed. Transformation does not necessarily make entirely new relations of care or power. As we see here, transformative practices can exacerbate already existing relations of power.

After Barbara Harrisson won her battle with Eve, Eve and the other orangutan infant Bob learned to live with the Harrissons in their feral garden. Eve slept in a cage built by Bidai, who cushioned it with a nest made of blankets, wood shavings, and leaves. Like Bob, Eve ate a daily diet of fruit and milk thickened with rice.[15] She learned to walk. Physiologically, orangutans are adapted for climbing through trees, not walking on the ground (MacKinnon 1974). Harrisson records how this happened:

> Eve was taught to walk. Bidai or I put her down on the lawn and walked away ten yards. She usually screamed madly, feeling abandoned, then (still screaming) started scrambling towards us. We then moved slowly away from her, keeping a distance. Soon she was so occupied with the task of walking towards us and controlling her limbs, that she forgot to scream. After a while she walked quite well and started to explore the garden on her own if we sat down with her. But she never ventured far; if we were out of sight for a moment, she always started screaming. (1987: 42)

The threat of losing bodily contact was powerful enough for Eve, whose body was not evolved to walk, to learn how to do so. In the absence of explanation as to why they sought such a skill from Eve, readers can guess that walking was one example of the exploration of species living together, of living out a feral future in their co-presence.

The description of orangutans learning to walk is given within the narrative in a section about exploration. In it, Bob learns how to interact and engage with the Harrissons' free-range turkey named Pauline. Exploration teaches him the taste, texture, and smell of pineapple bushes, the fruit it bore, and the insects that burrowed nearby. Here, exploration is central to the experiment and beyond the control of a human investigator and scientist. Perhaps it is more than coincidence that the exploration Harrisson encourages among her charges is reminiscent of scientific exploration, one of the key German traditions of knowledge production, as quintessentially demonstrated in the travels and encounters with strange and yet familiar others in Alexander von Humboldt's *Cosmos* (1868). Constrained within Harrisson's garden at Bunglo Segu, encounters across species are among the few possibilities open to them.

Ape motherhood came not only through inquiry in the controlled laboratory, regardless of its makeshift qualities in the domestic space of Bunglo Segu, but also through the explorations of fieldwork. Harrisson led a Sarawak Museum field expedition, which included the reluctant Gaun, the Sarawak Museum's most competent collector, taxidermist, and ornithologist, to observe orangutans in the wild, near Gaun's ancestral home in Sebuyau. She writes, "I knew that Gaun was not too keen to go with me. I sympathized with him in a way, but it was too late to change my sex" (1987: 54). Harrisson, writing about events in 1957 and first publishing in 1962, makes a joke of how time and convenience curtail her possibility of becoming transsexual. The joke is as much about the then-new technological makings of sex as it is about the ways in which her own gender positions her as an outsider to biological field research.[16]

Harrisson understood her idea of ape motherhood against the possibilities opened up by Dayang's sympathy and the consultations Harrisson made with human mothers about infant care. Ape motherhood for Harrisson was not about surrogacy and tenderness. Rather, it was about instilling independence through following a more "toughened approach" than she had in the first months of caring for infant orangutans. Harrisson wrote: "The trip toughened my approach. I made up my mind to try and be a true ape-mother from now on. The children would have to spend much time in the trees instead of on the ground. Perhaps they could learn to build themselves nests. Anyway, they should live as free of me as could practically be" (1987: 91). Harrisson's understanding of ape motherhood meant providing less interaction between herself and her charges. Being a human ape mother meant letting the infants go to the treetops and other places beyond human access, hoping for their potential nest-building instincts to appear on their own. Perhaps most importantly, it meant being distant from Harrisson.

How could infant orangutans learn to become orangutans without their mothers, while being as free as possible from Barbara Harrisson's interventions? The responsibility fell on Bidai. Bidai was the son of Nimbun, a friend of the Harrissons who was a Grand Master Tukan of a large Selakau longhouse. Harrisson explained the arrangement: "Nimbun is a distinguished friend of ours, and when he asked us to take his son away from the remote long-house, to see and learn a little in town, we naturally agreed" (1987: 36). Bidai brought his childhood friend Ina with him. At the time, Bidai only spoke one language and had to learn Malay, the language spoken across ethnicities in Sarawak's capital city Kuching. He and Ina earned a wage through the Sarawak Museum and learned how to skin and label specimens; however, Bidai in particular proved adept at tending to live animals.[17] Bidai used the wages he earned from the Sarawak Museum to "buy smart shorts (or, better still, longs), shirts and tie, a radio," and they both aspired to "become government servants with a regular salary . . . to have friends of different races and backgrounds, to go to night school; and for the holidays to go back to their village, big men" (1987: 37). The project of hosting Bidai in the Harrissons' home was a project of colonial intervention, a personal education in modern life.[18] That modern life paradoxically included teaching displaced orangutans their nature.

After Harrisson had her epiphany about what she called "ape motherhood," she called upon Bidai to enact it:

> "From now on," I said to Bidai, "put them into these trees every afternoon. Don't bring them down, even if they climb high. Let them do as they please."
>
> He did not like the idea and protested: what were we to do if they tried to get away? I told him not to worry. We must see for ourselves, I said.
>
> Eve was frightful at first. Every time Bidai put her in a tree she came down again to be cuddled. I teased him:
>
> "You have spoiled her too much! Look; she behaves like a tiny baby, she cannot be without you!"
>
> He was indignant. Had he not looked after her well always? Given her food, leaves to play with every day? Cleaned her cage, bathed her, never ever smacked her?
>
> "She must learn to grow up Bidai, live in her own world, get away from babyhood. Don't allow her to cling to you, teach her to stay in the trees. Sit up there with her. Perhaps she will get used to it then!"
>
> So he climbed half-way up the tree, and she went too, first clinging to him and then finding her own way. (1987: 91–92)

Harrisson's ape motherhood required Bidai's hands-on work and close attention to Eve. In that respect, it resembles the work of care in the colonial bourgeois household described by Ann Stoler, in which the banalities of upholding ideas of cleanliness produce the disciplining of colonial bodies and the formation of their subjectivities (Stoler 1995; Stoler and Strassler 2000). Yet the work of care is not meant to produce or maintain bourgeois European purity, but to foster a space for native subjects to find their own way at the behest of the colonial authority attempting to distance herself.

Some readers may be tempted to identify the animality shared by Eve and Bidai, who both climb the tree at the request of the British colonial in their midst, Barbara Harrisson.[19] Yet such a reading would be misplaced. It would require a disdainful regard for tree climbing as animal behavior, when it is a skill cultivated in Sarawak. It would be more productive to think instead about ape motherhood's expectation of finding one's own way, an expectation shared between Bidai as a Sarawakian youth and Eve as a Sarawakian animal, as if that way for each individual is natural, when the experience is entirely new for both. Such an expectation resonates with the hope for decolonization as projected by the evacuating force: in the aftermath of transformative contact, those who are expected to stay are also expected to become independent, but only in the terms and spaces set forth by the one bestowing that independence from afar.

The purpose of ape motherhood was to instill independence. However, the success of the orangutans' achievement of a semi-wild state in the Harrissons' garden led to an impasse: "Both the comforts and the confinement of our garden at Pig Lane were opposed to carrying the experiment further. What was now required were *less* comforts and a real *territory* sufficiently large to incite roaming; and an adequate food supply within. Unless we were able to fulfill these conditions, our Orangs would now develop away from half-wildness towards domesticity and eventually become zoo animals" (1987: 110, original emphasis). Territory is a loaded term that emphasizes the politics of land, since only those in power have the sovereign right to delineate territory. Although she was a self-identified "British colonial," she did not have real colonial authority to seize land and claim a territory. The patch of trees and pineapple bushes growing on the property on which she and her husband dwelled were not enough to satisfy the curiosities, interests, and needs of the orangutans growing under her care. Swaths of tree branches had already been destroyed from the orangutans' explorations. The material constraints of lacking territory were all too real.

Harrisson in 1959 entertained the possibilities of what could happen if resources could be marshaled to develop a sanctuary in which these displaced orangutans would be able to roam freely and be "liberated" (1987: 111). However, she felt that true liberation would be impossible: the orangutans would be so acclimated to humans that they would be susceptible to getting captured and illegally sold. Harrisson mused that they "might easily end up, after liberation, in a third-class zoo in Manila or Johor" (1987: 111). Harrisson came to learn that orangutans learn their behaviors: "from man in civilized surroundings like ours he [the liberated orangutan] adopts other humanized standards which eventually make him a zoo animal" (1987: 111). Contact with humans has thus impressed upon the orphaned infant orangutan a life that is too human to return to the wild, a life that is too transformed by "civilization" to be liberated.

Unable to provide adequate material conditions for a feral or independent life, Barbara Harrisson transferred the orangutans under her care to what she saw as the best solution: modern zoos. This led her to what she called a dreadful dilemma: "How can anyone who has loved animals like these bear to put them back behind bars? . . . All of our feelings are in favour of avoiding sending them into a zoo, even if excellent. This is one of the great heart-breaks in rearing large and slowly maturing animals, rescued from orphanage, damage or starvation. The dilemma is dreadful" (1987: 113). She then recalls how Europeans in Sarawak "have to" send their children to boarding schools in England (1987: 113).[20] The cruelty of separation was justifiable for Harrisson when it was best for the juvenile's future prospects.

Yet human caretakers *not* separating themselves from young orangutans would also be cruel. Barbara Harrisson suggests this when she meets another "ape mother" from Kuching's Chinese merchant class around April 1960. An infant orangutan had come under the care of the Harrissons right at the moment they were handling the travel arrangements to Edinburgh Zoo for the three infant orangutans under their care at the time. The report of the new orangutan's confiscation was too brief for Barbara Harrisson's liking, but it did contain the address, name, and sex of the illegal owner, Madam Goh. Harrisson went to her house and waited to speak to her so that she could get a better grasp of the orangutan's life history.

By visiting the former caretaker of this illegally possessed orangutan in order to gain more information, Harrisson exercised a colonial authority accessed through her position within the colonial state bureaucracy as one who is privy to the reports issued by the Forest Department. Yet she refused the assumption that she represented the policing colonial state. Rather, she represented care and the desire for the continuity of the same kind of care that the former caretaker offered, "so that there will be no mistakes."

Madam Goh explained the extent to which she cared for this infant in terms of love by carrying him around "all over the place" and by feeding him milk, rice, rice porridge, cookies, and oranges "like to a baby" (1987: 134). Harrisson at that moment thought to herself that this lady "had made use of the little animal to relieve a lot of her repressed feelings, poor dear" (1987: 134). Harrisson seems to pity her conversation partner as one who unknowingly mistreated this infant orangutan by imposing her sense of a warm universal motherhood that could melt divisions across species.

However, such tender motherhood would violently face its limits as the orangutan matured.

In her conversation with Madam Goh, Harrisson represented her authority as a scientifically oriented caregiver. She gained this authority through her experience of having raised many orangutan infants to health: "I consoled her by telling her that she would have been unable to keep the animal for long in a home like hers. That the baby would have doubled his weight and strength, his ability to climb and destroy things within less than a year. 'I have had many Orang babies,' I told her, 'and they get big and strong quickly. You cannot keep them as pets for long without being cruel to them, even if you love and feed them well!'" (1987: 135).

Harrisson positioned herself as a consoling advisor based on her authoritative experience of having hand-reared many infant orangutans. Yet just a moment before, she had positioned herself as one who needed counsel in order to obtain more information on the life history of the infant and the precise qualities of care that the infant received. She did not exercise her authority to reprimand the woman for not offering correct care. Rather, she exercised an authority of knowledge sympathetically to try to explain the futility of loving an infant orangutan, who will ultimately become strong. Continuing human love for an orangutan will result in the cruelty of captivity.

The encounter with a universal, interspecific tender motherhood against her own ideas of ape motherhood posed an ambivalent question for Harrisson. Without her conversation partner knowing it, Harrisson was taken to a difficult turn in which she personally had to confront the question of who has the right to be an ape mother and whether she, Harrisson herself, could truly be one: "I told her about the zoos where many animals are kept by people who understand them and treat them well and that I would find a good place for her baby. But when I came home and looked at my four charges I suddenly felt guilty and utterly wretched" (1987: 135).

Harrisson hints at the inability to be truly sure that she can provide the best care. Stretched by the difficult task of handling four infants at once, constantly worried about finding adequate places for them in updated modern zoos, she may have doubted herself at that moment. Although she carried herself as a British colonial and had the colonial power of information and surveillance, she ultimately did not have the colonial power to gain territory for her charges.

Readers are left with an awareness that she alone is the legitimate authority over all orphaned orangutans in the state of Sarawak, that she and the museum employees working for her are solely authorized to care for these creatures. She and her workers have a monopoly on caring for displaced orangutan infants and any possible joy to be obtained from that interaction, yet they alone have the impossible task of providing care that is invariably inadequate. Throughout the monograph, readers see that there are not enough trees to properly exercise the infants and that the only future homes for these animals are modern European and American zoos.[21]

Reflecting on the situation nearly fifty years later, Barbara Harrisson explained that her efforts were doomed to fail because Sarawak Museum's efforts were consistently underfunded by the colonial state: "what they [the colonial state] did was to put wildlife control under the museum, rather than forestry. But, the museum really didn't have any money to do this properly. So in colonial times, earlier times, the problem then was very small. But that is really why I got in touch with orangutans because the Museum had to confiscate orphaned animals seen in private possession or in trade. And they [the orangutans] were taken over by the Museum and somehow they had to be looked after. That is everything."

We can gather that underfunding led to the ambiguity between private and public care. Without her interventions, the apes would have been left in captivity on the museum grounds until their death. She speaks of herself as neither hero nor savior, but rather simply as part of an organization tasked with the responsibility of responding to the threat of pending extinction while being denied the resources to do so.[22]

Fifty Years Later

The colonial era of human–orangutan relations headed by Barbara Harrisson left material traces in the creaky wood-paneled images inside the old Sarawak Museum, sandwiched between dioramas of taxidermied orangutans and proboscis monkeys, jammed into the manila folders of the museum's back-office photography department, and boxed in the steel file boxes of the museum's archive. Bunglo Segu is in disrepair and is now fenced off. Once the administration of the state's wildlife was transferred from the Sarawak Museum to the Forestry Department in the early 1970s, the museum's previous role in rehabilitation was of little interest to the new administrators. What Barbara Harrisson had described in 1965 as a "long tradition of co-operation" between the conservator of the forest and the Sarawak Museum was replaced with brief memos about demolishing the remnants of the first and failed orangutan rehabilitation project in Sarawak's Bako National Park.[23]

Illustrating the disconnect, an officer at Batu Wildlife Center in May 2010 became agitated with me when I privately brought to his attention that the first rehabilitation efforts in the state began in the 1960s, a decade earlier than the time that he had announced in a welcoming address to the public. He raised his voice and said it was "bullshit," that people like to lie about their accomplishments in reports and monographs to make themselves look good. The only reference I had heard or seen anyone make to Harrisson's work was when Lin would turn to the appendix of Harrisson's monograph *Orang-Utan*. She found it useful since the appendix charts orangutan age and dental eruption. With it, she was able to estimate the ages of the new infant orangutans that came into the center's care. Harrisson's holistic descriptions of colonial Kuching, such as visits upriver to Iban country in order to study orangutans in their natural habitat, nighttime discussions with Iban mothers about their Chinese pottery heirlooms, and camouflaged viewings of orangutans while patiently waiting for hours high in a tree, all appeared to be of little interest to Lin, Ben, Layang, or anyone in today's Sarawak. Indeed, when Cindy adopted a similar narrative style for her first report on orangutan training for the Forest Corporation, Lin told her to rewrite and resubmit it with a more objective and scientific tone.

Although Harrisson's narrative may seem to bear little relevance to the lives of present-day custodians, her understanding of how to cultivate relations of rejection between orphaned orangutan and human caretaker are valid to this day. Photographic images produced under Barbara Harrisson's leadership are still relevant to how current custodians articulate a vision for orangutan rehabilitation today and in the years to come, along with relations of rejection now practiced as "tough love." The images of the past help facilitate a vision of a future that harkens back to aspirations from the past. In this "history of the future," orangutans are to be independent, despite their need for people (Koselleck 1985; Rosenberg and Harding 2005).[24] Today's workers at orangutan rehabilitation centers face the same crucial problems that Barbara Harrisson faced: How to sustain infant orangutan life without an orangutan mother? How can infant orangutans learn to become orangutans with the help of humans? These questions are mediated through caregivers' gendered and racialized bodies.

Tough Love

It was near quitting time when Ben asked me how it went when I accompanied Apai Julai and Gas for rehabilitation training. We both knew that Gas, a four-year-old juvenile orangutan, was particularly drawn to women and sought physical proximity whenever possible. I told him that I hope she habituates to my presence.

Ben said to me: "You need to be tough." Ben was my slightly younger peer and an urban Bidayuh. Since he graduated from the local college with his degree in biology four years previously, he had been working for the British volunteer company that supports orangutan rehabilitation efforts in Sarawak, Malaysia. Ben continued with his advice: "Bring a stick next time. Use it." I'm sure Ben could see the apprehension on my face. I couldn't bear the idea of threatening Gas with violence, let alone enacting it by hitting her with a stick. I was too deeply schooled in ideologies of positive reinforcement, having previously worked as an intern zookeeper in the United States. As we parted, he said: "You have to be tough."

Gas was at a crucial time in her life, in which two possibilities were open to her. If she were to be deemed too human-imprinted, she would be kept permanently captive and part of the breeding program, in which she'd be occasionally caged in with a male orangutan until she likely became pregnant. Because of the sexually dimorphic differences in size and strength between males and females, she would be unable to escape his attempts at copulation within an enclosure built for just one. If she were to be deemed "semi-wild" and rehabilitated, then she could have a free-ranging life within the confines of the wildlife center. Hence, the investment in tough love was an investment in some kind of autonomy for her in the future.[25]

Gas's attachment to human women was traced to her early childhood. She came in at the age of two from an oil palm plantation. The previous volunteer program manager and his girlfriend cuddled with her around the clock for almost a year. Their acts of poor judgment included bringing her to the public market as they did their own food shopping. They were fired shortly thereafter. Taking Gas home was one of many breaches in human–animal intimacy at the wildlife center during their time. The orangutans are supposed to be either in cages or outside, not cuddling in human arms. Still, almost three years after their departure, whenever a woman was present, Gas would stare, and if possible she'd reach out to touch her, especially if she were outside Gas's cage.

To be a caring animal keeper at a rehabilitation center was to reject tender human–animal intimacy. Wild orangutans try to avoid human interactions, while semi-wild or rehabilitant orangutans try to engage human interactions. Thus the work of a caretaker was paradoxical: an animal keeper needed to keep orangutans away from humans as a human himself.

Calls for toughness were set in opposition to "softness." Keepers explained that being soft was a failure to be tough and would result in vulnerability to an attack by an orangutan. Specific persons were pegged as likely targets. From June 2009 until May 2010, Ching and her baby Dylan were free-ranging in the forests of the wildlife center. She had a history of attacking people, specifically women. I talked to Lin about it. She was younger than Ben, but second in command at the wildlife center. I asked her whether it mattered if the women Ching chased were local or foreign:

LIN: No, it doesn't matter. As long as it's a female. I've been chased by her. Rachel [Tom's girlfriend and coworker] and volunteers [mostly white] have been chased by Ching as well. Even Apai Len. Apai Len. Because according to them [the keepers], . . . Apai Len is a bit soft.

JUNO: He is soft?

LIN: Yeah, he is soft. Umm, you know. A woman is soft normally. And so when Apai Len is a bit, how do you say, he's very soft and his behavior, his attitude is more like a lady, something like that. So that's why this orangutan is also chasing Apai Len as well. (March 25, 2010 [21:02])

This attitude toward "softness" directly criticized femininity and effete masculinity. Apai Len was a grandfather. Yet his effete way of talking, walking, and sitting warranted the pejorative description of "womanly softness" from his fellow orangutan keepers.[26]

By explaining that Ching attacks women and effeminate men, the work of caring for her and other semi-wild apes was coded as a manly profession, and this coding was attributed to Ching's behavior and agency. Doing so obscures the male keepers' own investments in the masculinity of their labor, yet their investments in their manliness resurfaced in jokes about who was perceived to be womanly by an orangutan.[27]

Ben's boss, Tom, was convinced that Gas was especially fixated on blond women, since the girlfriend of the former ENGAGE project manager was blond. Yet Lin, Cindy, and I would disagree. Lin and Cindy were the junior officers at the wildlife center. None of us was blond, all of us separately accompanied rehab training, and all of us experienced Gas's attempts at making physical contact.

Gas's strategy for intimacy seemed fairly consistent, yet nevertheless always surprising in its suddenness. From up in the trees, Gas would stare down at the keepers and myself or the junior officers observing training. She would play-fight with Lee, a male orangutan two years younger, and then suddenly tumble down and grab hold of me, at which Apai Julai would get up with a stick in hand, and if Gas didn't leave my side immediately, he would whack her. Or she would be at the tree canopy. To move, she would use her body weight to sway the trees, and then she'd suddenly sway down and fall on me. Or she'd be on a tree, and rapidly descend to the ground and appear to knuckle-walk toward the keepers, but then make a sharp turn toward me and grab hold of my hand and lick it, at which Apai Julai would rush toward her with stick in hand and try to whack her. If she didn't get back up to the trees before he got there, he would hit her until she did.

When Gas tried to approach me, I would try to turn my body away, my back to her and arms to myself. I would dodge her calculated falls and sprint to a position between the two keepers. Cindy said that she purposely avoided eye contact and constantly turned her back. Her form of engagement was attempted disengagement. However, it was a futile effort. Gas's constant return and Apai Julai's constant hits served as reminders that disengagement was impossible. Co-presence already implicated me in the dynamic of punishment. Indeed, *our* co-presence was the reason for the punishment.

I decided to stop attending their trainings. I had already obtained enough notes, and being the constant witness to and thereby participant in everyday violence no longer felt worth it for me. As a researcher, I was able to leave. But for those unable to leave, for those who are directly responsible for orangutan rehabilitation, what alternatives do they have? Must they use punishment to exercise their custodianship? Is using pain and the threat of inflicting pain the only way to engage and respond to the apes under their care?

Faced with these options, Lin decided to douse her hands with eucalyptus oil so that if Gas approached her, she would recoil from the sting. As Lin herself said to me, "She'll hate you after that." Cindy continued her method of disengagement. Two months later, when training Gas and the other orangutans together, the inability to disengage was viscerally brought to everyone's attention: an orangutan sank her teeth into Cindy's hand two centimeters from her tendon, almost permanently disabling her.

For those staying, the choices were tough love, smothering softness, or the two possibilities demonstrated by Lin and Cindy. Like Lin, you could embody tough love not by using beatings and physical strength, but with the help of pain-inducing agents. Or, like Cindy, you could subject yourself to pain and potential debilitation in the name of rehabilitation. Every alternative was fraught with ambivalence because everyone knew what would happen to Gas and others like her if their rehabilitation attempts failed: deemed too human-imprinted, she would be kept captive, unable and unwilling to live like an orangutan, while at the same time unable yet apparently wanting to be close to humans.

Docility

Both Gas and her predecessor Eve fifty years earlier faced the same problem: taken from their orangutan mothers and raised by humans from an early age, they sought out but were denied human affection. They shared the same dilemma of how to be cared for as orangutans without their orangutan mothers. The people who cared for them saw that the root of the problem was in the young age at which they had been forced to live with people. Thus the problem was in their docility. Docility was visible in photographs from Barbara Harrisson's experiment.

The image of a vulnerable infant orangutan with pleading eyes has become a cliché for orangutan rehabilitation in coffee-table books, fund-raising websites, and the genre known as animal docu-soaps (Candea 2010; Payne and Prudente 2008). Yet the image evoked the same feelings from the animal handlers Layang and Ricky, staff officers Syed and Lin, and every volunteer I interviewed: dislike and disapproval. How was it that so many different custodians, with their different kinds of stakes, experiences, and interest in orangutan rehabilitation, had the same kind of feelings about this figure?

Syed was a man of much experience. When he was young and just entering the workforce in the early 1970s, he wanted to become a police officer or work in immigration. Like Bidai in the 1950s, Syed had wanted a good, stable job, and a government job was just that. He was able to become a civil servant by working in the forestry department. In the Parks and Biodiversity and Conservation division of that department, he worked his way up from worker to warden and then to officer. He later became the head administrator of Batu Wildlife Center and eventually took control of Lundu Wildlife as its "person in charge," the highest officer in command of a site.

Upon seeing figure 1.1, Syed said that putting clothes on an orangutan would be duplicating humanness. He suspected that as it got older, the orangutan would become accustomed to wearing clothes. He said, "treating it like a human" is dangerous "because they are too close to you in certain ways." His fear is that orangutans have the intelligence or "mentality" to be able to adopt human-like behaviors that cannot be enacted in the future, such as wearing clothes or smoking cigarettes or other behaviors rehabilitant orangutans might pick up while being reared in human households. The duplication of human behaviors by orangutans or chimpanzees, for Syed, was better suited for a zoo than for a rehabilitation center.

Syed said he "totally disagreed" with the photo, and not just because the infant in question was an orangutan. Even if the infant was a macaque or gibbon, he strongly "still disagreed." He said, "Yeah, very good, but what about when they are a certain age when they cannot be controlled?" Growth, and with it added strength, is inevitable. They become a danger: "to not only you, but to others." Syed continued: "I totally disagree with people keeping wildlife, as an expert. Because from experience, these animals are wildlife, growing up. And actually they have their own natural behavior at a certain age. If they want to mate, if they're breeding lights come up, that will not be from proper care; their mentality changes. They become aggressive. It's normal for wildlife. . . . Sooner or later, you abandon them because of this threatening problem" (January 9, 2009).

When seeing a still image of an infant, Syed saw the inevitability of bodily growth and changed behaviors. The image was not frozen in time for Syed, but represented a promise for the future, specifically that infants eventually become mature.

Upon seeing the image of Eve cuddling Bidai, figure 1.2, Layang saw her inevitable dependency due to the young age of the orangutan: "She grabs him like the mother. If this orangutan [is] in the wild . . . they can be

FIGURE 1.1 An infant orangutan cared for by Barbara Harrisson and Bidai anak Pengulu Nimbun. Image reprinted with permission from the Sarawak Museum.

dependent. Because this person, they keep it, when they are still young. So this orangutan doesn't have a choice. They must follow this man" (January 4, 2009). Dependency for Layang was marked by a lack of choice. If the orangutan were older, she would be able to choose whether or not she wanted affection or intimacy.

After our interview, Layang lingered over the photo. He wanted to place the time of the photo. I told him it had been taken in the 1950s. When I asked him what he found so interesting in the image, he sniggered and told me to look. After a moment of silence, he entertained my question:

FIGURE I.2 Bidai anak Pengulu Nimbun with Eve at Bunglo Segu. He sports the latest hairstyle. Image reprinted with permission from the Sarawak Museum.

Because the human doesn't use a T-shirt. Because he's staying by the—what we call it—the since over fifty years before—British, ya, when still British. . . . Because we, like these people, they stay in the longhouse, they are far away from what we call the city, so when they try to go the city, they take a few days. Sometimes a week. Or a month. Going to town, to buy things. . . . Mostly, these people [are] what we call still traditionals, you know, using the traditional way. And even also, it's difficult to find the money, doing that kind of thing. What they [are] just doing is [*sic*] the activities in the longhouse, always

> during that time [when] still a colony: hunting, going to the river, planting *padi*. Only that is their jobs. So they don't have a construction job. It's not like now. Now you ask the young man to stay in the longhouse, the young man doesn't want.

The image spoke to Layang about the disjuncture between the present and past conditions for indigenous people of Sarawak. Layang himself was born in the city, but his father was a man like Bidai, pictured in this photo. Like Bidai, Layang's father came to the city of Kuching from upriver, near Kapit on the Rajang River, back when it would take about a week to get to the capital city using water routes. Layang may not have noticed how Bidai's hair was slick with coconut oil pomade, styled in a ducktail, combed up and defying gravity in what was the new urban rock 'n' roll fashion. Layang could not have guessed that Bidai's first salary went toward purchasing a transistor radio. Layang did not comment on Bidai's hair as a sign of urban modernity. Layang instead seemed to fixate on Bidai's bare torso, which served as an index to an un-modern past and unevenly developed present, one in which work involved only agrarian labor for the self and the immediate community, and not wage labor in a cash economy. Living life independently in the longhouse was a relic of the past, according to Layang. The present circumstances of wage labor, of dependence on money, were the means of independence in Layang's world.

Conclusion

Dominating others for the sake of the well-being and survival of those being dominated, working assiduously for others' freedom, but lacking the power to delineate territory that would make that possible within the spatial framework of the colonial state, and unable to remove the influence one has upon the other, we can see that Barbara Harrisson's "ape motherhood" was an unrealized project of decolonization, contemporaneous with formal decolonizing efforts in Sarawak. The juxtaposition of a decolonizing Sarawakian past with a postcolonial present points to the ways in which an aspiration for "the future" harkens back to the past. Looking to the past, we get a glimpse of relations across species, from the unruly cultivars growing at Bunglo Segu to the unconventional modes of care developed through experimentation. Looking to the present, we get a glimpse of the violent attempts at forging distance between humans and displaced orangutans undergoing rehabilitation. Both the past and the present work toward the same future: one where orangutans can be taught by humans to be independent of humans.

The idioms of ape motherhood and tough love give us a sense of how to understand the cultivation of a feral future for semi-wild orangutans. Both are idioms of care that culminate in rejection. This is the work of care—labor that is performed with or without monetary compensation and embodied through the signs of gender and race.[28]

Today's tough love is rooted in a disavowed colonial history of ape motherhood—even in its shared investment in the production of science from the periphery and in experiments in independence that result in a feral life. Their disavowal is particularly rooted in the way they gender certain forms of care as feminine, namely motherhood and care taking place in the home. The disavowal of femininity is especially expressed in the idea that feminine "softness" leads to orangutan attacks. Such disavowal occurs despite Barbara Harrisson's transgressions against white colonial femininity entailed in her ape motherhood that was "toughened" by field research. Those transgressions took the form of humor around the idea of getting a sex change in advance of a wild orangutan expedition, the flourishing of a feral home antithetical to colonial bourgeois society, and the transformations of her colonial home into a colonial laboratory through the use of a pipette for force-feeding.

When facing the possibility of an orangutan dying in your company, what would you do? Would you intervene by force-feeding, as Barbara Harrison did? Or would you let the orangutan die, perhaps in recognition of the orangutan's agency, as Dayang conveyed? The men training the orangutans, in a way, perform rehabilitation as acts of forced feeding: the cruelty of injuring an orangutan is undertaken for the sake of their survival and their semi-wild state. The women working as officers appear to try an alternative, a third way akin to force-feeding but with less force, making them survive and thrive with a little less violence in their lives. I wonder about a fourth way: What would it be like to come to terms with Eve's death and the death of her species? The next chapter follows yet another kind of intervention: commercial volunteers who intervene in the lives of orangutans by giving money and their own manual labor.

Chapter 2

ON THE SURFACE OF SKIN AND EARTH

Through the thick cover of leaves, the Iban worker Ngalih and the two British commercial volunteers Liz and Kate spotted the Bornean orangutans Ching and Baby Dylan. The view of an orangutan mother with her clinging infant up in the tree canopy was the reason the volunteers had traveled from England to Sarawak, each paying two thousand dollars to work there for two weeks—excluding the cost of airfare. The very idea of this encounter motivated their sweaty labor the day before as they hauled planks of wood from one area of the park to another. And now, spontaneously, they were in the moment of being at that interface.[1]

For Ngalih, who was fairly new to the job, having worked there for less than a year, this was yet another test—one he encountered every time he spotted Ching when he was the sole worker around. All the keepers knew that semi-wild orangutans, even juveniles, could injure people seriously. He could not be friendly with her since they didn't have a history together, like the one she had with Ricky, Ngalih's coworker and fellow subcontracted worker, who had worked at the orangutan rehabilitation center since Ching was transported to the site in 1999.[2] Nor would it be wise for him to convey confidence like Layang, another fellow subcontractor who had worked with orangutans since 1991 and was thus experienced with the semiotic, material exchanges of human–rehabilitant orangutan encounters. At that point in time, Ngalih's experience of the shared space inhabited with semi-wild orangutans was nearly as fresh as the orangutan baby's.

Ching then did exactly what Ngalih had hoped would *not* happen: she started climbing down the tree. Everyone at the park knew of her history of attacking people, especially women. Ngalih told the women to run and hide. They did so and successfully evaded the possibility of bodily harm.

This moment gives us a hint about the complexity of encounters while working at Lundu Wildlife Center. Oscillating between elation and fear within a single, spontaneous moment, affective feeling happens in the space between bodies, on the surface of skin (Massumi 2002). These sensations do not just take place in space, where space is merely a backdrop to the action (Massey 1994, 2005). Rather, where these sensations occur matters for the relations that develop and dissipate between bodies. Geographic space, with its long history of specific relations, make any present relation possible. When the British volunteers who have paid to be there are guided by an Iban in a forest that is now a national park and they spot a displaced orangutan, they brush against each other, against histories of land development, as they are embedded in colonial legacies and millions of years of evolutionary departures between hominids—specifically, orangutans endemic to Borneo since the Pleistocene and humans who have since the Pleistocene made their way throughout the world.

In this chapter, I consider how relations with others always entail a collapsing of time between the seconds and microseconds of encounter and the years, decades, and even millennia over which differences between bodies are generated.[3] Such efforts at relating to one another do not just reinforce distinction (Choy 2011). They also reinforce a momentary arrangement or "agencement" (Phillips 2006; Roelvink et al. 2015). In other words, we are different and together; what I have become is only possible because of what you have become, and the encounter between us had and has the power to transform each of us. This is the matter of affect, which we feel on the surface of skin and on the surface of earth.

By paying attention to the space between bodies, we see how affective encounters mediated through the rush of sensations that are felt on the surface of the skin and on a specific surface of the Earth generate both individual subjects in relation to one another and a global economy. The particular economy generated here, in this contact zone, attempts to be different from other modes of exchange and exploitation.[4] People from wealthy countries pay to perform challenging manual labor. Yet old colonial, capitalist, and human inequalities are very difficult to escape.

Whereas anthropologist Eduardo Kohn (2013) is interested in the general patterns that can be gleaned when we study relations and cultural forms, I am interested in the surprises and unpredictability that unfold in specific situations, the moments before they become lived experiences that could set possible patterns in the future. Indeed, the moments I am interested in might never duplicate themselves; rather, they happen because of contingent circumstances. Stories involving Ching, Deh, and other orangutans in Sarawak's wildlife centers show how unpredictability is contingent on the moment in which different actors subject themselves to contact with others and how the unpredictability of having been there at that place and at that time drives the billion-dollar-plus commercial volunteerism industry (McGehee 2014). Surprising possibilities, which are completely dependent on co-presence, are not replaceable with charity or the technological mediation of a real-time webcam or "crittercam" footage (Haraway 2008; Kapoor 2012; Parreñas 2016). To feel that they are part of something, volunteers have to be there in the flesh, to feel things on the surface of their skin and on a specific surface of earth—Borneo, Sarawak, Lundu Wildlife Center.[5]

This chapter examines the everyday custodial work of care across potentially incommensurable differences. Sweeping out cages and construction work was the ordinary care work at Lundu Wildlife Center, but under specific circumstances, it could become extraordinary. This was apparent from the moment I began shadowing Layang at his job, performing "animal husbandry" in the morning. Husbandry here consisted of shoveling scat, sending orangutans out to their enclosures, and feeding them.[6] I myself felt the extraordinary qualities of the everyday interface when I performed a small part of his job, which entailed carrying a three-year-old orangutan from a cell to an enclosure because the gate connecting the two had rusted beyond repair. Commercial volunteers gladly engaged the potential vulnerability of being at the direct interface with orangutans, even if their work forbade the physical sensation of touching an orangutan. They cleaned cages, built construction projects like pathways and fences, and created enrichment parcels for orangutans to destroy. As I show in this chapter, they enjoyed participating in manual labor at the site, feeling vulnerable at the interface of extinction and doing *something* about the problem of species loss by making *some things*.

The stories here share an underlying tension in which the lack of certitude about what happens next or what could happen convey how the work of care, as a compensated activity for a low-wage worker and as an activity that is seen as a commodity by vacationing ecotourists, is about feeling mutual vulnerability. Yet that vulnerability is not evenly shared across the interface. As much as commercial volunteers would like to live out non-exploitative forms of exchange, their participation always ended in their eventual departure. They, more than anybody at the park, were most able to evade bodily harm.

The Work of Care

Of the nine new orangutans arriving between 2006 and 2010, four came from human settlements bordering oil palm estates that were converted from forest, including the juvenile female Gas. Two came from the illegal personal menagerie of an alleged gangster specializing in illegal timber.[7] Along with the adult dominant male orangutan Efran, these confiscated and surrendered orangutans were among the thirteen orangutans living at Lundu Wildlife that I regularly encountered while conducting field research.[8]

The Forestry Corporation took these animals into custody on behalf of the state of Sarawak and in accordance with the state's Wildlife Ordinance. Orangutan care is primarily custodial in that it is about maintaining their facilities and their welfare, which requires manual labor. This kind of custodianship is different from but related to stewardship, which is rooted in the British manorial system, when someone took the place of an absent landlord to husband the resources of an estate (Hainsworth 1992; Hughes 2006a).[9] Custodianship here is more along the lines of the North American sense of custodial care: physically demanding and humble work that gives one an intimate knowledge of others and their space.

To understand the relations custodians and the animals in their care develop with each other, I turned to the keeper Layang. Layang's reputation had already preceded him before we met. Every person, from Forestry Corporation officers like Lin to the British volunteer company project manager Tom, swore that Layang commanded respect from the orangutans and other animals. I was told that he was able to get the very large, formerly free-ranging, and flanged adult male orangutan James to get into a cage without resorting to sedation. Layang was as small as Lin or myself, not much more than five feet (150 cm) tall, yet he would sometimes stand inside the enclosure with James and directly hand him food. Layang and James encountered each other every day and were thus acclimated to each other's physical proximity.

I regularly accompanied Layang on his morning animal husbandry routine, which meant getting all the orangutans out from the night den to their exhibits, cleaning their cages, and feeding them. On my first morning observing this, Layang tested me with the juvenile orangutan Gas, who was three years old at the time.

Layang knew that when it came to orangutans, I had read a lot but had no experience at the time of my arrival. My experience as an intern at an American zoo handling siamangs and a sun bear did not count for him. Layang likely assumed that I would come in and pose as an expert—as many Forestry Corporation staff members with college degrees had done. After all, how could learning from books teach one how to do the job of handling orangutans and of dealing with the affective exchange between person and orangutan?

Layang said to me, "Bring her."[10] It didn't sound right to me. I knew that volunteers were strictly forbidden from gaining intimate contact with orangutans, yet I was not to be there for just a month or two weeks. I also knew that the door from her cage to the enclosure was broken and that someone would have to carry her out.[11] I asked, in disbelief, "You mean go into the exhibit and carry her?" Layang said, "Yeah. Just carry her and take her."

Communication might be linear in that it is directed toward a future response (Ingold 2007; Kohn 2013). Yet communication between Gas and me was as linear as a freely drawn squiggle; by no means was bodily communication between Gas and me the shortest distance between points.[12]

> She [Gas] grabs my hand, but held on to the posts inside her night den with her feet—refusing to go, so my grip of her belly slips to below her chin and she then slips down because my hold slackened, because I was afraid of choking her. She's on the floor now and grabs hold of a rope dangling between us. She uses me to get a better angle on the rope

> and then she starts holding my leg with both legs. She then nibbles at the zipper pocket of my pants. I start to walk towards the gate and she then bares her teeth and is about to bear down on my leg. I'm a little scared since every one of her teeth is bigger than mine. I try to hold down my panic and I raise her arms and try to angle her away from me, but she twists in an awkward way. . . . I tried to figure out how to move with her and she perhaps did the same. (field notes, November 18, 2008)

Calling this encounter between bodies "choreography" would be a misnomer (Thompson 2005). Carrying a juvenile orangutan was a far from graceful experience in which the movements of our bodies were intentional. This was an interface of confusion. I was surprised by the coarseness of her hair, what felt to be the unusual distribution of her weight, and the sheer awkwardness of trying to walk with another heavy body, one not physiologically equipped to walk, but rather to climb (Rodman and Cant 1984). Gas gripped my arm, leg, and hand with four different muscular and clammy hands as we walked, sometimes dragging a limb or grasping the bars and walls as we passed. The texture and surface of her skin and hair interfaced with my skin and clothes. In that interface of bodies, our bodies responded to one another, but in ways that were asymmetrical, unequal, and muddled. The job of carrying Gas to her enclosure entailed a hyperawareness of both our bodies in the fleeting moment of encounter. Cooperation did not come on its own, and neither could it be physically forced. Though she was little, she was still stronger than I was.

The affective surge between us, along with the mutual vulnerability inherent in the risk of me hurting her or her hurting me, was at the core of custodial labor. Vulnerability for each of us was in the possibility of feeling pain. It was also in the shared possibility of sharing microbes or parasitic worms. However, that risk was not mutually shared. At the largest rehabilitation center in Malaysian Borneo, a longitudinal study using data from 1967 to 2004 found infant mortality among semi-wild orangutans to be 50 percent more than among wild populations (Kuze et al. 2008). This was likely due to anthroponotic illness and the close proximity with other orangutans that would readily transmit illnesses. Humans are vectors of illness for orangutans, and most humans who come into proximity are either unaware of this or cannot be derailed from their tourism plans: 15 percent of 633 surveyed visitors over nine days in 2007 self-reported being sick with either sore throat, congestion, fever, diarrhea, or vomiting (Muehlenbein et al. 2010). It goes without saying that this figure does not include the visitors who were subclinical carriers of any number of infectious diseases that could be transmitted to the orangutans for whom they cared enough to see in the flesh.

ENGAGE tried to mitigate anthroponotic risk amid the H1N1 pandemic of 2009 by obliging everyone to wear surgical masks in the sweltering heat. Wearing surgical masks served as a technoscientific ritual to ward off something fundamental about the proximity semi-wild, semi-captive orangutans like Gas had with her caretakers: living together entails sharing microbes.[13]

My bodily encounter with Gas happened before the H1N1 crisis. Neither Kuze's (2008) nor Muehlenbein's (2010) research was known to any of us. I would have been one of the few to be able to access their research findings, considering research journal paywalls and the time it would take to read and digest such articles. Once we got to her enclosure, moving her away from me was harder than getting her out of her cage in the first place. I responded to her four grabbing limbs by pulling her to my torso and holding her at my hip; letting her limbs wrap around my torso was not the smartest move. Although I vaguely remember removing her forearms and swinging her, it would have been possible to detach myself only if she herself were ready to let go. I was a bad keeper because I took too long for what was considered a simple task, and I responded with affectionate touch instead of what the custodians called "being tough."

Encounters like these constituted a small but essential part of the job of being a worker at the center. Custodians had to manage their time to make sure they could limit the intensity of these encounters to the space of two hours while cleaning up ten dirty cages. While Gas had my attention during the moment of interface, I did not have much time to reflect on the work of co-constituting the affective space between my human body and Gas's orangutan body. There were six more orangutans to bring out, including ten-year-old Lisbet and three-year-old Mut. As for the latter aspect of my shortcomings, I was to find out later that my response should have been more "tough" and less affectionate.

By having me plunge into Gas's night cage with no experience and with little instruction other than the guidance of a few imperatives, Layang showed me how he had to learn his job from day one. He recalled that day with me once when we sat down at the veranda outside his home for a formal interview. He said that it was 1991 and he was seventeen when he started at Batu Wildlife Center, the parent site of Lundu. He was handed a broom and a dustpan and told to clean up scat inside a cage that at that moment housed eight adult orangutans. If I thought a bite from an infant orangutan was nerve-wracking, I clearly could never have survived Layang's early career. Layang was tough and gladly demonstrated it.

After I carried Gas out to her enclosure on that first morning of my participant observation in the orangutan night den, Lisbet and Mut, together in their shared cage, were not going out of their own volition. It was 9 AM and the sun was already out, baking the concrete walls of the exterior exhibit, heating it like an oven. Layang yelled, "*Keluar!*" ("get out") and jabbed the broomstick against the railing. Lisbet just moved around the ropes above while Mut sat on the floor. Layang then took the fire extinguisher and pointed the nozzle at Lisbet while he repeated his imperative. Lisbet had Mut holding onto her as she climbed to the highest ropes, away from Layang and the fire extinguisher. This time, Layang took out the water hose and started dousing Lisbet and Mut. By then, Mut was on a platform inside the cage squealing, while Lisbet was above, unsuccessfully dodging the spraying water (MacKinnon 1974).[14] Once Layang turned the water pressure off, Lisbet climbed up the ropes again near Layang and urinated and defecated. At that, Layang told me to leave the night den. From where I stood outside the building, I could see him continue with the hose. Soon he gave up and stormed out to the quarantine to get the blow gun. Upon seeing the blow gun, Lisbet ran out with Mut clinging to her.

The act I witnessed, regardless of whether it could be construed as violence, was a deliberate cultivation of an affective relationship. Layang's method of cleaning that day reminds us that the affect in affective relations should not be misread as "affection" in the sense of tenderness. His labor as an orangutan keeper requires a heightened sensitivity to bodies in motion and co-presence (particularly his, theirs, and mine) as well as the flexibility to respond. His job of custodial labor is to hone and respond to the sensations produced through the interface between him and his orangutan charges and to face the risks inherent in such an affective encounter. As he shows, a custodian's response entails rejecting demands for affection and tenderness. He as an animal keeper needs to keep orangutans away from humans as a human himself. Truly caring for an orangutan means responding to the intensity between bodies while also rejecting the other's efforts to gain closeness, affection, or a different direction of movement.

While others implied that there was something innate about Layang's skill, he himself explained that it was a matter of experience. He cultivated his relationships with the orangutans under his watch. He regularly maintained physical contact with the orangutans. Ching, even when she was pregnant, let him regularly rest a gentle hand on her belly. He would play with Lisbet by entering her cage, throwing his hands up like an orangutan, hulking toward her, and posturing his walk in the way that she walked.[15] She was bigger than he was and could have clobbered him if she wished; yet it never happened. These moments of joy, play, and risk also constituted his custodial labor.

More than a year later, Layang and I talked once more about what it takes to be an animal handler at a rehabilitation site. One of our friends was bitten during an orangutan jungle skills overnight training session; the wound from the bite became infected and took six months to heal. Bites were not surprising to Layang. It was part of the job. Layang had explained rehabilitation to me: "We need to teach them while they are still young to be afraid of humans. . . . But if they're like Ching, you ask them to be afraid of humans, we [handlers] cannot [force them be afraid of humans]" (interview, May 27, 2010). This was a rare moment in which Layang admitted to a degree of futility in his work. Rehabilitation is set up with a perfect orangutan in mind: young yet independent, one already and continuously fearful of humans. Yet if the orangutan is young, how can she not be impressionable? As the orangutan experiences more interactions with humans, including being fed by them, how can she remain fearful of them? How can the orangutan not help but be enculturated through co-presence with those around her?

Everyone knew that Ching's aptness to fearlessly engage humans and attack them was a result of her proximity to humans from the time that she was two years old. As Layang explained to me, the only person she is ever afraid of is the handler: "she is afraid of the person who is the handler. The rest. . . ." His face turned sour and indifferent, completing the sentence for him. He continued, "They will give respect to that person who feeds them, only that person she will respect. But sometimes they change their mood. That person can be in what we call not in a safe place. That's why you need to read their mind."

I asked, "How do you read their mind?"

He responded plainly, "You need to see their face. If they have a hot temper or not." Reading an orangutan's mind meant reading the orangutan's body and sensing affect generated between orangutan and custodian. Faces were not the only vehicles of expression for orangutans (Tempelmann et al. 2011). Hair standing on end invariably conveyed anger to anyone regularly in contact with semi-wild orangutans, at both Lundu and Batu Wildlife Centers.

Hair standing on end is subtler than the physical display of male adult "hair shaking" that primatologist John MacKinnon (1974: 61) describes in one of the first studies on orangutans to systematically examine orangutan gestures and vocalizations. It is also more subtle than the "poke" and "pursed lips" Tempelmann et al. (2011: 436) describe in their experiment testing attentiveness among orangutans and other great apes begging experimenters for food. Since custodians and rehabilitant orangutans come into close physical proximity to each other nearly every day over periods of years—closer than the distance experienced by the primatologist standing on muddy ground looking up into the canopy of trees and without the mesh barriers that separate animal subjects from human researchers within the confines of the laboratory—their interface consists of cautious exchanges and the production of feelings and possibilities in which no one is in control. Yet workers like Layang must attempt to enact the feelings of tough love and risk being bitten while the orangutans are in a position where they need keepers like Layang to survive.

Layang shared an informed personal experience of human–orangutan interface through conversations. With only an ability to witness signs, perceive feelings, and speak to people, how do we understand the other side of this interface? What can we call the intensity that makes an orangutan's hair stand up? The handlers and local managers are certain that it is anger. But what is anger when experienced and conveyed by an orangutan facing a human? Calling it an emotion leads us into the circuitous pathways of animal cognition that often privilege the underpinnings of logical rules, semantic function, and hierarchies of intelligence in communication (Hauser et al. 2002; Miles 1993; Russon and Andrews 2010).[16] Yet this hair-raising feeling is something communicated, palpable, and produced in the interface between bodies—in this case, between a human custodian and a rehabilitant orangutan.

How would a human know if they are correctly or incorrectly understanding what is being communicated? Is there always a correct answer? Was Layang right in using the hose as a weapon? I suspect that there are no correct answers. Instead, there are responses that beget other responses. Layang's employers expected him as a custodian to respond within this interface—without getting seriously injured. This is the risk that Layang had to face every day at work. The affective sensations here are both uncertain and productive. Its uncertainty is evident in the risk of potentially debilitating bites, and its productivity is evident in the attentiveness gained by keepers and orangutans.

Commercial Volunteerism

Lundu Wildlife Center needed both workers and volunteers to keep running as an orangutan rehabilitation center. Revenues from volunteers' fees and the outcomes of their volunteered labor fueled the orangutan rehabilitation center. Volunteers usually stayed for one month for about four thousand dollars. The minimal amount of time one could stay as a paying volunteer was two weeks, which was introduced in response to the global financial crisis of the late 2000s. They need at least six monthlong volunteers every month to stay afloat. Volunteer revenues were used toward supplementing the animals' diet, building supplies for structures that have a hard time withstanding rot in the tropical humidity and destruction by curious and playful orangutans, and improving the infrastructure of walkways that connect the site's various enclosures.

Forms of intimacy or physical proximity in the interface between human visitors and orangutans at the site posed serious potential dangers. The orangutans could contract human-carried illnesses, and the volunteers could be subject to bodily harm. Even though the volunteers were not directly handling orangutans, they cleaned their cages in close proximity to the animals. They had to pay attention to "grab zones" that were within reach of some orangutans. Their co-presence with the orangutans also entailed a need to be aware of the interface between themselves and the orangutans in their midst. The potential vulnerability of volunteers to bodily violence was especially felt one day with the orangutan Efran.

Efran's body size and musky smell were formidable. His weight hovered around two hundred pounds, and he stood less than five feet tall. The strength of his arms could bear the weight of his entire body. His hands were perhaps three times the size of his handlers'. His face, with fully developed cheek pads, was about four times the size of my own. Efran was a fully grown, twenty-three-year-old adult male. As an adult male orangutan, he was perhaps ten times stronger than any of us humans at the site. Like the other orangutans at the site, Efran as an infant had been displaced by habitat destruction and was confiscated by the state as an illegally trafficked pet.

When newly arrived volunteers at the orangutan rehabilitation site first see Efran, they are often struck by his presence and the multiple sensations evoked by the encounter. Nearly everyone would gasp, especially when meeting his gaze. Some would cringe at the smell. Others would enjoy the fleshy odor. Taking in the presence of such a rare proximity, one volunteer said at that moment of encounter, "Wow . . . moments like these make me feel like I'm really here." The evocation of multiple forms of senses called for a heightened awareness of their co-presence in the interface. It was through affective encounters that volunteers felt that they were really there.[17]

Co-presence produces feelings that are impossible to mediate through sight and sound alone. New video technologies including crittercams can mediate connections and intimate knowledge, but they cannot mediate the driving force for commercial volunteerism involving wildlife: the possibilities for direct bodily engagement and intimate encounters afforded by co-presence with an animal endemic to a specific area.[18] The ubiquity of technologically mediated forms of intimacy adds to the value of real, in the flesh co-presence, such that volunteers treat the hard labor necessary in wildlife rehabilitation efforts as a commodity to be consumed.

After years of failed efforts at rehabilitation, Efran served as a warning to volunteers against being too affectionate toward a young orangutan, which could contribute to the classification of being unfit for rehabilitation. When Efran was a free-ranging subadult at Batu Wildlife Center, he was too habituated to people and became too dangerous: he hand-walked on the ground and attacked people by biting them.[19] At Lundu, Efran was among the orangutans held in enclosures during the day for what local managers had explained were "educational and conservation purposes." Part of the orangutan husbandry routine included releasing Efran into his enclosure and cleaning his night den. He would regularly linger near the gate, gripping the gate's heavy steel bars and watching from outside while a volunteer cleaned inside. He would often open his mouth and wait until the volunteer using the hose understood his gesture as wanting a drink from the spout (Liebal et al. 2006). Sometimes he would put his hands under the running soapy water and would rub his palms, to the delight of the volunteer washing his cage. As long as he was on the other side of the bars, everything was fine.

One morning in 2008, Efran was using a bamboo stick to pile together and sweep the debris of fruit rinds and his own feces out of his cage to the gutter of the interior walkway, imitating the keepers' and volunteers' cleaning motions, as he was occasionally apt to do. To get Efran to leave the cage, that morning's keeper would have to first put fresh food out in his enclosure and then leave the cage door to the enclosure open. Since Efran preferred staying inside, leaving his food in the enclosure was a way to coerce him to go out. Once he went for his food, a worker would close and lock the cage door behind him, and then two of the volunteers could clean his cage.

On that morning, the volunteer Eva was in the corridor just outside the cage, nearly at the door. Fay was inside the cage, giving it a last rinse. Len substituted for Layang that day and stood near Eva. Eva saw that Efran could open the cage door about ten to twelve inches. Right after the incident, Eva and Fay explained to me what had happened:

EVA: I noticed something was not right. He could push it open more than usually. It all happened so quickly. I think I told Len, "I think he can open the door." I'm not sure if there was a language barrier or something. So then Len lifted it [the locking mechanism] up, right while Efran was playing with the door then. I think I yelled, "he's coming inside!" And Fay quickly threw the hose out, jumped out, and slammed the [heavy, wrought iron] gate behind her. So then Len acted quickly enough to secure that door shut. . . . Now it's scary, now that nothing happened. Had something happened, we'd have a different story!

FAY: It was so scary. We should *always* make sure to check it [the lock] twice. Thank you, Eva, for saving my life.

EVA: I didn't think it was saving your life; it was just the moment.[20]

Eva and Fay had the luxury of speculation, since thankfully nothing tragic happened. However, it would be a mistake to think they were speaking in hyperbole, since the threat of bodily injury was very real. When seeing Efran, it is impossible to forget his strength. As soon as

Efran returned to his cage, he slammed the dangling oversized truck tire across the room, conveying what everyone—including Layang, who heard about the episode after the fact—assumed was anger. Anyone who had read Biruté Galdikas's (1995) memoir would have remembered her story of witnessing a rehabilitant orangutan forcibly copulate a worker at Camp Leakey.[21] Considering Efran's earlier record of biting people while he was free-ranging, it was likely that he would feel free to bite anyone once free from his confines.

Although not tragic, something did happen in the interface between Efran, Eva, Fay, and Len. What happened was not just a matter of gathering one's wits quickly; it was also about being attuned to the production of affect in the interface. The intensity was palpable for all four of them. The volunteers and Len needed to be attuned to the moment, to Efran, and to each other. Such a moment of surprise and the actions that followed do not follow a general pattern, but create new possibilities of relation. In this case, Fay, Eva, and Len's response was about the prevented potential terror of too closely encountering a being with which one would not normally be in close contact outside a wildlife center like Lundu. Fay and Eva's experience helps show how intersubjective sensation, affectively charged in a complex, momentary feeling between terror, pleasure, and relief, drives the encounters that occur at this site. Embodied intensity, or affect, is made sharper through encountering wild animals endemic to Borneo. This interest held by volunteers can play out here precisely because the moment is fleeting.

The Gendering of Affective "Hard Labor"

Helping with orangutan husbandry takes about a quarter to a third of a volunteer's laboring day—just as it does for workers like Layang. The rest of the time is usually for construction projects requiring what a few volunteers had described to me as "hard labor." Volunteers' hard labor is also crucial to the maintenance and development of the center. Their work includes such tasks as hauling wood hundreds of yards from the work yard to the construction site, lifting thirty-pound river stones and moving them in order to support the foundations of the orangutan ranger station, removing trees felled by heavy rain, using simple workshop tools to extract metal beams buried into the ground so that a better enclosure can be built for the gibbons, as well as sawing and varnishing wooden planks to make a fence to block visitors from entering or disturbing the orangutan jungle skills training site.

In this regard, performing hard labor is a commodity consumed by vacationing postindustrial workers. It personifies a relationship between the postindustrial Global North and the industrializing Global South where performing southern labor can become northern leisure. Hard labor at the wildlife center also entails an affective interface. The interface is not limited to the attunement between lively bodies, but also includes inanimate forms at the surfaces in which bodies meet. Affect in this respect is produced through the toiling labor of one's gendered body in relation to other gendered bodies at a worksite and the heightened awareness of one's own body and others' through hard labor: heat, sweat, and muscular pain. Affect in this context goes through an extra step of attunement: the intensity is not only for the individual pleasure or joy of inhabiting and pushing one's body through duress, but for the welfare and future of the animals at the site, for what volunteers see as an improvement in the plight of endangered animals. This is where we can see how intersubjective affect has greater significance beyond the subjects interfacing with each other.

In this regard, the affect of hard labor becomes a "sentiment" or "affective idea" of custodial labor for volunteers (Yanagisako 2002). In this particular case, affect is given a conscious purpose and meaning. Volunteers like Muriel imbue this interface with personal significance, which in turn resonates for a demographic of postindustrial workers that is large enough to produce a tourist economy of commercial volunteering.

In all their activities entailing hard labor, the volunteers I encountered were aware of the masculine undercurrents of their labor, in which physical brawn was needed to get the various jobs done.[22] Yet most of the volunteers are white, professional women, mostly from Britain and Australia. Of the approximately 120 volunteers I encountered, only about ten were men. How did these women participate in custodial labor, and how did they come to terms with the masculinized forms of this work?

The volunteer Muriel's experience and her narrative about it help answer this question. Muriel and a co-volunteer participated in the effort to remove a street lamppost that was blocking the way for a future gibbon enclosure. They had worked under the powerful heat of the dry season sun for a couple of hours already. The lamppost was unnecessary because the electricity was never fully wired and no one was allowed to come into the park at night. They, along with Layang, another keeper Ngalih, and Tom—the British volunteer company project manager—managed to dislodge and knock down the lamppost without power tools, just rope, shovels, and their sheer physical strength. Wire circuits dangled from the bottom of the lamppost. Layang and then Tom took turns trying to cut the wires using bolt cutters. They couldn't make a dent. Muriel suggested that they open the wire box of the lamppost. Ngalih and Layang, talking among themselves in a mix of Iban and English, suggested using the handsaw. While Ngalih left to get the saw, Muriel repeated herself. It sometimes took a couple of tries for others to understand her Scottish accent. Layang heeded Muriel's suggestion. He opened the wire box and found that there were no visible wires—just a plywood backing, with nothing in it. We all laughed as Layang reached his arm in and pulled out the wires from the box—they had been loose the whole time. I could not resist a quip, "Sometimes it takes more than just brute strength!" Tom responded, "We didn't use all of our brute strength capacities—we could have gotten angry at it and kicked it!"

What Tom and I had jokingly called "brute strength" referenced the sweaty, muscle-straining, backbreaking work of intensely inhabiting one's body, in the presence of other similarly toiling bodies. With their bodies in contact with each other and facing the resistance of an incredibly heavy object, affect surges between human bodies and between human body and a physical body, specifically a lamppost. While affect here nearly erupted into the emotion of anger, it simmered in the heat between their sweaty bodies and fatigued muscles and continued to motivate them.

At this site, the strength of one's hard labor is prized by male workers and is a commoditized fetish for the female volunteers. By paying to engage in this work, Muriel—who works at a major insurance company at home—and other postindustrial, white-collar women like her—embodied that power herself. What is to be gained in the purchase of embodying manly muscle power? Muriel offered an answer when I asked her if she wondered why so many women agree to volunteer their hard labor:

> No, because I know women are more likely to do this because women I think are kinder and patient and they're more likely to volunteer. And I think women care more about animals than men do. I don't know whether it's a maternal thing, they want to nurture automati-

cally. And I also think that other women prove to themselves that they can actually be as good as a man, come all this way [to Sarawak], and do *hard* work. And some of the women have families, children, and I think they just want to do something on their own and get a sense of self-worth, self-esteem, because as it is four [PM], you work all day and come back dirty and tired, you know you worked just as hard as a man. So I'm not surprised that it's all women. I'd be more surprised to see young boys here. You're more likely to get men, like Tony, who was older, retired, and had time or men like Jack who got dragged here by their girlfriends. I can't imagine a man, like a man my age, just who works in an office same as me, I can't imagine him volunteering to do it, because I don't think a man could care enough to do it. I think women just generally care more.[23]

Muriel's answer offered two interpretations. On the one hand, she implied that engaging in physically demanding labor is about gender equality, that a woman could work just as hard as a man, even in labor forms that are skewed along the physical differences of gendered bodies. On the other hand, Muriel resignifies physical, hard labor as a form of feminized care work. By resignifying hard labor as a form of care labor, perhaps Muriel was allaying a gender anxiety reminiscent of the anxieties directed toward Victorian colliery women in Britain who developed masculine embodiments while working as laborers (McClintock 1995). By emphasizing our present location as distant from her home in Scotland, she seemed to recall the trope of colonial white women at the frontier of empire (Burton 1994; Jeffrey 1998; Mills 1993). Although I may not be able to access her unspoken thoughts, it is clear that Muriel seeks to define care to an essentialized femininity of wanting to nurture. Thus, she understands the hard labor that she performs at the site as a form of nurturing and caring. Yet the work of volunteering at Lundu Wildlife, as demonstrated by Eva, Fay, and Muriel, has very little nurturing in it, and neither does the work of rehabilitation as performed by Layang.

The care in Muriel's labor is not the care of emotional labor or traditional ideas of care work (Hochschild 1983; Parreñas 2001b). She is not there to personally nurture a baby orangutan. Rather, care for her is conveyed through the fact of paying to work and working very hard. The work of hard labor is prompted by the care and interest for the future of another species. While Muriel sees this aspect of her volunteering effort as perhaps a womanly interest in the welfare of an animal, I see it as a form of custodial labor. Muriel participates in custodial labor by toiling to produce the material effects of improved welfare for the animals at Lundu Wildlife and by paying to perform that labor. She laboriously and financially gives with the intention of gaining a certain kind of future for others, specifically endangered orangutans that can potentially be rehabilitated.

Donating funds for environmental charity is not enough for Muriel and other commercial volunteers who pay to perform custodial labor. Charity would deny them the experience of engaging affect while in the proximity to rare wildlife and while engaging in toil.[24] When I asked her about her occupation, her response exemplified how her own and other volunteers' participation has to be understood in a postindustrial context:

JUNO: So then what kind of work do you do back home?

MURIEL: I am in an office, a computer, a big insurance company, boring, sitting, nothing, no value to anyone. It's just a big company, making money out of people, for insurance and pensions. And *no* self-satisfaction in my job. I hate it. And just wanted to get out and do hard work for a change. But of course, you can't just pack it in and do it full time, because obviously at the end of the day, you got to go back. And a secure job is a good job, so that's why you do this for a month. At least, you might only do this once, but it's still a month that you've done *something*. So it's good. It's good and it'll be hard to go back and to sit at a desk and look at silly things that are really quite pointless, when you know that something like this is happening.

Muriel's frustration and lack of satisfaction with her job puts a finger on the pulse of the current political-economic moment. In Muriel's scathing critique of her workplace and the insurance industry, she points to the neoliberal moment of corporate wealth, diminishing pensions and benefits, job insecurity, and social precarity (Allison 2013; Berlant 2011). Muriel, like others in the Global North, are deeply alienated from the products of their labor in the service economy. Muriel turns to commodifying manual labor and paying to participate in meaningful production. Doing "something" meant producing material products or "some things" by one's hard labor. It meant phenomenologically engaging in the world through intensely inhabiting one's body, being affected by other bodies—living and otherwise. It meant personally responding to the perceived need to improve the conditions of endangered animals of a postcolony vis-à-vis her own manual labor and paying to perform it.[25] This is where custodial labor comes to have meaning for Muriel and others like her.

Conclusion

Ultimately, Fay, Eva, Muriel, and the other volunteers all left after their month was up. The experiences for which they paid were worth it: they felt affect, experienced risks, and all safely returned home to Britain. Efran, Gas, and Lisbet all likely continue to be excited behind their bars when staring at and engaging with new volunteers, as they had done every month when I was in their midst. Even though Layang complained about the Forest Corporation's futile attempts at becoming profitable and even though he still had the contacts to pick up construction work, he stayed with this job that entailed affect and risk every day until it got the best of him. When on the job trying to rescue stranded wildlife near Bakun Dam in January 2011, he contracted a lethal case of melioidosis and died within a week.

The custodial labor performed by Layang and the commercial volunteers points to the ways in which affect produced between bodies also produces a global economy through a dynamic, interspecies interface. This particular economy is inflected with postindustrial desires for meaningful labor, embodied toil, and affective interfacing with endangered wildlife as well as the conditions of postcolonial inequalities that expose some to greater risk than others. Understanding trans-specific care and affect characterizing it requires understanding how everybody in relation to another is vulnerable to the other, and that mutual vulnerability entails risks and consequences that are unequally experienced. In the next chapter, we will see how female orangutans face specific risk and vulnerability because of their sex and how volunteering women come to terms with such violence.

PART II **ENCLOSURES**

Chapter 3

FORCED COPULATION FOR CONSERVATION

By 4 PM, workers at Batu Wildlife Center would already have directed visitors back to the parking lot or to the road that leads to the bus stop at the main entrance. By 5 PM, it is quitting time, when all the humans leave the orangutans to their own devices. Once in May 2010, during that final workday hour, I stood in the courtyard as I heard grunts high over my head in the canopy. I walked a few steps further down the path toward the vacant veterinarian bungalow so that I could get a better look. I could see the female orangutan nicknamed Grandma. She earned the appellation when one of her progeny had an offspring. Grandma herself had recently given birth. Her infant was clutched tightly to her torso. Up in the tree above me, she was with the subadult male Aqil, who had recently turned eight years of age and become sexually active. He was attempting to hold her limbs down while facing her. She vocalized and he remained quiet. They wrestled limbs, holding position for seconds at a time. She positioned herself behind him and turned her body away, so that her back was turned to his back. He repeatedly grabbed hold of her limbs with his limbs and manipulated their position so that she would face him instead. While holding otherwise still, with three of his hands locked onto her wrists, he placed his free hand near her genitals and smelled his fingers. He repeated this action. After another minute of wrestling and her grunting, he stopped and directed his attention to the bananas below that had been abandoned on the platform.

In the midst of watching this, Peter, the person in charge of Batu Wildlife Center approached me:[1]

PETER: Did you see them mate?
JUNO: Uhh . . . yeah. . . . [hesitant]
PETER: She's not ready; she still has a little one.
JUNO: What do you do when this happens?
PETER: Nothing! We don't interfere. It's nature. The only reason why we'd interfere is if the baby is getting injured and then we'd yell and try to stop.

Writing this now, I doubt copulation happened in my presence. In retrospect, I think I witnessed a failed attempt. Yet forced copulation was the subject of our conversation, even though we did not describe the act in that way.

Peter's words conveyed an idea of nature that required forgetting or ignoring the conditions that brought these orangutans to this place. It required ignoring that we were standing in a built environment, paved with concrete, surrounded by aerial electric wire, and staffed with ten workers beneath his command. It also required ignoring how the space of the center was carved from 15 km^2 to 6.5 km^2 within the span of fifteen years, between 1995 and 2010 (Chow 1996). It required overlooking the possibility that such diminished space would have an impact on the orangutans' social lives and behaviors.

For the orangutans on site, life at the wildlife center meant living together instead of in relative solitude. It meant acclimation to humans. It meant the facilitation of reproduction without regard to the context in which reproduction occurs. By thinking of the consequences of rescue, I wish to show how saving members of a species from extinction inflicts new forms of violence. Could rescuing wildlife facing extinction bring about a life worse than death?

Whether you are compelled to answer yes or no, thinking through this question has us imagining levels of harm and forces a standard answer to the question of what makes any life bearable. Instead of imagining gradations of harm, which would be an oversimplification for the sake of getting a standard measure of comparison, let us think of the consequences of human interventions on the spaces and lives of displaced orangutans. For people intervening, what forms of violence are permissible at the level of individuals for the good of a species' future?

We can understand this dilemma as a question of scale between individual and species. I address this question by thinking about the physical space of Sarawak's two orangutan rehabilitation centers. In this chapter, I examine the relationship between anthropogenic space and human encouragement of orangutan sexual reproduction. These two issues produce a system of sexual violence, where violence is experienced between individuals for the sake of producing future generations of an endangered species. This is where "reproductive futurism" meets "compulsory heterosexuality." What kind of life is deemed worth living, when the survival of a few individual members of endangered species is at stake, when their lives come to stand in for the entire species? How are these kinds of lives determined, and by whom?

Queer theorist Lee Edelman's (2004) analysis of "reproductive futurism" in American cultural politics is useful here in Sarawak. Reproductive futurism for Edelman names the rhetoric that uses the idea of future generations as a tool for political mobilization and action (Seymour 2013). In the way that the figure of the child represents the clean future and is a powerful symbol for both mainstream gay politics and fundamentalist Christian politics—sides usually positioned at odds with each other—so too is the symbol of the infant orangutan for Forestry Corporation officers, commercial volunteers from the Global North, and others concerned about the pending extinction of orangutans.[2] Efforts to encourage sexual reproduction among displaced orangutans compel the question: At what social and psychic cost are future generations of orangutans produced, and who bears those costs? The answer lies in the material conditions of physical space constraints and orangutan bodies.

In respect to physiology, orangutans are sexually dimorphic, such that females are smaller and have shorter reach than males. The geographic boundaries of Batu Wildlife Center that were altered by land developers create a social structure in which orangutan females, outpaced by and smaller than males, have nowhere to evade social interactions and nowhere to gain the solitude that characterizes their behavior as a species—the most solitary of all great apes. The values decision makers place on heterosexual reproduction and the way anthropogenic space influences semi-wild orangutan behavior work together to create the conditions of "compulsory heterosexuality" (Haraway 2008; Rich 1981). Lesbian-feminist poet and theorist Adrienne Rich (1981) coined the term to describe the everyday human norms that aggressively push for a "male right" of access to women. The idea helps us understand how gender comes to matter for sexed animals engaged in the everyday struggle to find food and to find a living in a site of displacement.[3]

The physical space of Sarawak's two semi-wild orangutan sites shapes the ways in which orangutans experience social and sexual violence through their bodily sex. In this chapter, I explain how some Forestry Corporation executives, officers, and workers condone sexual forms of violence and intervene against other forms. By doing so, those supporting this social structure reinforce ideas of nature, sexuality, and wildness at the expense of the well-being of female members of the species.

Not all workers happily collude with these circumstances. Some, like the ranger Nadim at Batu Wildlife Center, empathize with their female charges and criticize the brutality of what he calls "rape" that happens on site.[4] Prominent primatologists, whom I think of as feminists, critique the usage of the word *rape* as a technical term in primatological literature. Thus, I turn to the debate between evolutionary biologists and anthropologists about the politics of the term *rape* and the extent to which acts at Sarawak's Wildlife Centers could be identified as such.

The decisions of the Forestry Corporation's management and the constraint of space at the center have led to the banality of forced copulation, a typical behavior observed among wild orangutans, but exacerbated among semi-wild populations. The officers and senior managers who are vested with the responsibility of how Sarawak's displaced orangutans are to live in Sarawak's sanctuaries are mostly men. Yet mostly British women volunteer to support orangutan livelihoods with their own bodily hard labor. How do these women come to terms with their support of structures of sexual violence? This question guides the conclusion to this chapter.

"The Ecosystem Is Dead"

For the Forestry Corporation, conservation meant supporting the reproduction of orangutans—regardless of the conditions under which future generations are born. The equation of conservation with reproduction was explicit at the November 2009 Regional Symposium on Orangutan Conservation facilitated by the Forestry Corporation management and sponsored by Sarawak state agencies. At that time, it was estimated that 2,500 wild orangutans remained within the state of Sarawak, while the world's entire population was estimated to be less than 50,000, with most orangutans to be found in Indonesia (Chan 2009). The famous orangutan primatologist Biruté Galdikas in her keynote address estimated that in a matter of about ten years, the small orangutan population in Sarawak will be larger than the population in Indonesia due to mass-scale conversion of forest into oil palm plantations. Palm oil is exported for vegetable oil, processed foods, and cosmetics in Europe, North America, India, and China. Galdikas took care in her statement to clarify that oil palm plantations were not in themselves detrimental to orangutan survival. Rather, she was of the opinion that industrial demands could be reconciled with wildlife conservation, that oil palm plantations and orangutans could coexist if a few adequate corridors for wild orangutans remain (Ancrenaz and Lackman-Ancrenaz 2004).

Inspired by Dr. Galdikas's remarks at the symposium, Datuk Len, who at the time bore two responsibilities as CEO of the Forestry Corporation and deputy permanent secretary of the Ministry of Planning and Resource Management, took it as an opportunity to express both his company's and the state's commitment to biodiversity. In the press conference following Galdikas's talk, he declared that the state government will "target to increase the orangutan population in Sarawak from 2,500 to 4,000" (Cheng 2009: 4). This was a surprise to the conference attendants. Clearly he had no idea about orangutan birth intervals, anatomy, or behavior. He was a successful businessman who saw exponential growth as good, for whom good assets were many assets.

An audience member asked if the Forestry Corporation planned to impose captive breeding. A senior manager of the Forestry Corporation took the microphone. He had decades of experience working in Parks and Biodiversity Conservation for the state and had the respect of NGO affiliates, coworkers in the Forestry Corporation, and his former colleagues in the Forest Department. He intervened to clarify, stating that there were still enough wild orangutans in the state to constitute a "healthy population" and that state conservation efforts to increase the species' population should focus on the conservation of their habitats rather than captive breeding. His opinion was that the area of Ulu Sebuyau should become a totally protected area, an opinion shared by the director of the Wildlife Conservation Society in Malaysia.[5]

The Forestry Corporation officer's intervention illustrated a tension in the organization's conservation mission between protecting wild orangutan populations on the one hand and promoting the reproduction of rehabilitant orangutan populations on the other. Despite his explanation that captive breeding is unnecessary at this time, the forms of captivity at both wildlife centers in Sarawak facilitate encounters in which male orangutans reproduce with female orangutans through forced copulation.

Efforts to encourage reproduction have been made at Lundu through a captive breeding program. In 2006, it was decided that the orangutan Ti should get pregnant. This site does not facilitate what biologists call "female choice" (Smuts 1987; Zuk 2002). Unlike zoos, neither site relies on artificial insemination. Instead, Ti was locked into the enclosure of the fully flanged adult male orangutan Efran. At the time, Efran was the only male.

Tom, who served as the field director of ENGAGE, retold the story to me and a group of volunteers:

> I've read about it in the literature, but man, was it brutal. The first time it was like, "Oh well; take it for the team, Ti." But for the next three to four days, he would drag her around the enclosure and she'd scream and eek so loudly and he wouldn't let go of her, even in sleep. She bit him, a minor bite, nothing serious. But for that he grabbed hold of her; pounded down on her head multiple times; copulated with her.

They removed Ti from Efran's dwelling early, out of fear for her safety. Yet this experience was not exceptional at the park. Three years earlier, the same treatment of locking a female orangutan in with the same male happened. That female orangutan, Wani, died shortly after her infant died at two weeks of age.

While Wani was held captive with Efran, someone took photographs of them. I encountered a color inkjet print of a photograph of that scene from 2006 on the billboard of a locally owned youth hostel that hosted commercial volunteers before and after their stay at Lundu Wildlife Center. The image was too poor to reprint and is more clearly rendered by way of illustration (see figure 3.1). The orangutans' names were printed together

FIGURE 3.1 Denying "female choice." Illustration by Tess Pugsley.

on the photograph—a parody of wedding announcements that were typical among newlyweds in Kuching. An untrained eye gets the impression that the image depicted a happy couple. On closer inspection, one sees that the male orangutan, much larger than the female, is fast asleep. The female in the photograph is wide awake and staring at the hand that grips her, even as he sleeps. The patches around her eyes are pale, signaling her remarkably young age.

Primatologists report that forced copulation is a typical behavior among orangutans (Knott et al. 2010). Yet what makes this behavior among rehabilitant orangutans particularly troubling is how forced sociality and lack of space contributes to the frequency and banality of forced copulation. This lack of space and the social relations that result from it are not found just at the site that looks like a zoo, but at the site that looks free. This site also does not facilitate female choice.

Batu Wildlife Center is the parent site of Lundu Wildlife Center. The original wildlife population at Lundu Wildlife Center all came from an overpopulated Batu Wildlife Center. With twenty-six male and female orangutans given free range within the confines of six and a half square kilometers, the space of Batu Wildlife Center facilitates far more interaction between orangutans than has been observed in the wild. For example, primatologist John Mitani's (1985) sixteen-month study at Mentoko in East Kalimantan, Indonesia (south of Sabah, Malaysia), in the 1980s observed 106 occasions in which males "associated" with females. Association meant encounters in which they were within thirty meters of one another. At Batu Wildlife Center, males and females associate with each other every day in even closer proximity.[6]

Orangutans across all subspecies are sexually dimorphic: males are much stronger and larger than females. When encountering female orangutans, even prepubescent females, subadult male orangutans would sometimes use their strength and force them to copulate. Batu Wildlife Center's wildlife ranger Nadim brought this up when I once asked him to help me identify the names of the subadult males newly approaching the feeding platform at which Nadim and I stood:

> Jeffrey. Morni. *Semua lalaki* [all males]. The others [other orangutans] I think are thinking, "I hate this place." *Binci buat* [hatred builds]. She comes just to eat. *They're free, but in fear. The ecosystem is dead.* It's not like in the wild, where they meet, male and female, in seven or eight years. Here, they meet every day! Deh, Grandma, *if they come here, they're forcing themselves to be raped.* You see that with Grandma and Deh. They come at other times now. *Their face so anxious.* They come early or late [in respect to feeding times]. They wait for males to not be there. Is Lucas there? Aqil, Ahmed, Jeffrey? . . . It [the problem that they meet every day, that there are too many males relative to females] was [a] fault from the beginning [since the park was founded]. Norma [he indicates the orangutan geneology chart right in front of him], in nature, three [offspring would be] already enough. [Norma has had 4.]

> *The ecosystem is dead.* Their babies are one, two [years of age]: baby still on Grandma, but she's being bothered by Ahmed, Jeffrey, Aqil. The males are happy. Ladies not. (original emphasis)

By sharing what he perceives to be Grandma's and Deh's feelings, Nadim shows a willingness to imagine the perspective of female orangutans. Nadim senses the female orangutans' anxieties and locates the feeling whenever he sees their faces. Picking up on the sensations that he perceives through the space of their bodies, in their shared interface, he imagines that they hate this place. This surfeit of feeling is part of what makes this place dead. However, the substantial reason for the death of this ecosystem, as Nadim suggests, is the proliferation of reproduction. Too much of one kind of life causes death. This is where compulsory heterosexuality, caused by the violence of diminishing space, becomes deadly. This is a dead ecosystem for Nadim, dead enough for him to repeat the weight of his words and let it sit in the humid air surrounding us.

The terror he perceives when looking at the faces of his female charges causes him to describe reproduction at Batu Wildlife Center as "rape." Yet by using a passive sentence construction, "forcing themselves to get raped," he evades blaming the "rapists." Rape for him references the coercion that results in sexual reproduction, but he does not call male orangutans rapists.

Evidence collected by primatologists in the wild supports Nadim's empirical claims about orangutan sexual behavior. In a study conducted among captive orangutans, primatologists Nadler and Collins (1991) found that when orangutans are contained in spaces in which female orangutans can exercise "female choice" through restricted access tests that allow their access to males but prevents males from accessing them, orangutans copulate at rates more typical of wild conditions. Furthermore, when female orangutans have choice about their sexual partners, they copulate less than in captive conditions where both males and females have "free access" to each other.[7]

Perhaps Nadim's words help explain such free access: to be free (bebas), but in fear. The experience of eight-year-old Sadamiah is just one such example. As an orangutan, she would reach sexual maturity at around the age of twelve to fourteen. One day, Sadamiah was "gang raped," according to Batu Wildlife Center animal keepers Boboy and Anggun. Up in the trees but still visible to attendants, fourteen-year-old Aqil and twelve-year-old

Ahmed pinned her down. Aqil held her down while Ahmed copulated with her and then the two males switched places. Many days after the incident, she still would move at an unusually slow pace. Getting food from the central feeding platform where she would encounter other orangutans became an even harder struggle than it already was for this small female orangutan, who, like the other female orangutans, would quickly exit once the older females Deh and Grandma entered the area or when any of the sexually mature male orangutans arrived.

The keeper Boboy suspected that the incident Sadamiah experienced with Aqil and Ahmed dislocated one of her bones necessary for arboreal locomotion, thus causing her to move slower. Nadim was not entirely sure; he went over possible reasons with me: "It could be that she's not ready to mate. It's like I told you, because they all meet at the feeding platform. It wouldn't happen like that in the wild. Maybe a few years and then you meet? Not every day! It can't be food poisoning, because if it was, all the orangutans would get it; or other orangutans would be sick" (interview, July 8, 2010).

Forced copulation was a potential cause for Sadamiah's slow movement and apparent hesitation to come to the main feeding platform to eat. As Nadim reminded me, it was not the first time we spoke about the disturbing encounters specifically experienced by female orangutans.

A similar incident happened when six-year-old Nini was wrestled, held, and ultimately forced to copulate by her brother, Ahmed. Two days after the incident, I overheard a tour guide explain the occurrence to two visitors. One of the visitors, whom I surmised to be a middle-aged man from either Singapore or peninsular Malaysia, replied, "I'm sure he wanted to because she's so beautiful." The woman in his company laughed.[8] The Bidayuh keeper Zeb responded: "Ahmed doesn't care that's his sister or his mother. We're trying to relocate them [Ahmed and Aqil] but it's not so easy." Zeb then shifted the subject to comparing human and orangutan behaviors: "In my culture, when two cousins marry each other, we tell them, 'oh you two are animals.'"

In their statements, forced copulation or "rape" among animals was naturalized by both the tourists and the keeper, but in different ways. The tourist's language both anthropomorphizes and trivializes prepubescent and apparently forced sex by connecting rape to sexual desirability. His statement resonates with the everyday misogynistic ways in which some speak about human rape in which the victim is blamed, in which the character of the assault survivor is doubted because of dress and behaviors (Bourke 2007; Reger 2014). The dynamic between him and his female companion, where he offers up a reason for sexual violence that is rooted in sexual desire for beauty and at which his female companion laughs, conveys how sexuality, especially nonhuman sexuality, is mediated through cultural concepts.

Zeb's statement, in comparison, emphasizes that Ahmed is an animal, one ruled by powerful instincts and not by his own volition or free will. Zeb seems to invite the visitors to imagine another way of making sense of what happens at the park. Ahmed as an orangutan is free of incest taboos and free of cultural constraints (Lévi-Strauss et al. 1969). Yet he is not free to act because something instinctive and beyond intention drives him. Sex, in Zeb's view, is purely natural for orangutans, despite the circumstances that bring Ahmed to be physically present and near his sister.

These encounters between the orangutans happen high in the trees. Because of the distance, workers are unable to intervene in these acts even if they tried. Yelling would not deter them. Tossed sticks could not reach them. Even though Peter claimed they would try to interfere if an infant was in danger, it would be to no avail. "Reproductive futurism" might inform Peter's prioritization of an infant, but the material constraints of the space, of the close proximity between orangutans, and the aerial distance often kept by orangutans choosing to stay up in the trees above people on the ground make it physically impossible for people to get orangutans to act as they wish. Such acts as the one involving Sadamiah are disturbing enough to be spoken about for days following any such event.

Violence was condoned only when it was sexual. Primatologists argue that violence between males is normal behavior (Watts et al. 2006). Yet Batu Wildlife Center, before and after privatization, including under Peter's leadership, would intervene when males fought and exercised natural behaviors. They would sedate the challenging male and transfer him to zoo-like captivity.[9] This was why both Efran and James were in captivity at the sister site, Lundu Wildlife Center. Zeb hinted at this future for Ahmed and Aqil. Finding a place for them at Lundu Wildlife Center, with its seven cages for nine orangutans, was already difficult. Making updates to the structure of Lundu Wildlife Center's buildings would be too costly and beyond the budget allocated to the Forestry Corporation's Parks and Biodiversity Conservation Division.

All the keepers point out that the problem is that there are simply too many male orangutans, all of whom presumably want to mate. When conversing with Nadim, I asked if the center would consider sterilizing some of the male orangutans.[10] To that idea, he sniggered in disbelief. "But they're endangered!" he replied. As endangered animals, their reproductive capacities are too symbolically potent. In this logic, the reason for endangerment is not deforestation and a shortage of space, but insufficient reproduction. Even though male orangutan sterilization could potentially enable female orangutans to live their lives with less fear, it would be at the expense of potential population growth. The potential alone for population growth ultimately trumped violence faced by female orangutans, even if it meant that sex involved prepubescent juvenile females and was therefore nonreproductive, and regardless of possible harm resulting from these sexual encounters.

Do Orangutans Rape?

Forced copulation? Or rape? What constitutes the acts identified by these terms? The primatologist Richard Wrangham (Wrangham and Peterson 1996), for instance, freely uses the term *rape* as a way to name sexual violence. Rape for Wrangham is one of many possible "reproductive strategies." Other evolutionary biologists, specifically Christine Drea and Kim Wallen (2003) as well as Cheryl Knott (Knott et al. 2010), decline the use of the word *rape* and call for more nuanced interpretations and analyses of what should be called "forced copulation." Here I turn to primatologists and evolutionary biologists. My intention in reading works by evolutionary scientists is not to find biological determinants or innate reasons for sexual violence. Rather, I am interested in the politics of semantics. Can we claim the word *rape* in the context of wildlife sanctuaries? Doing so would point to structures of sexual violence caused by facilitating compulsory heterosexuality for the sake of reproducing a future population of an endangered species. Could we talk about rape in such contexts without making arguments about the innate propensity of some to rape others and without extending liberal personhood to apes, especially when the political ideology of liberalism has little traction here?[11]

In *Demonic Males* (1996), Richard Wrangham with popular science writer Dale Peterson cites examples of forced copulation among orangutans as evidence of the evolutionary, primordial roots of human male violence. They define rape along the lines of controversial evolutionary psychologist Craig Palmer, who sees "rape as copulation where the victim resists to the best of her (or his) ability, or where a likely result of such resistance would be death or bodily harm" (Wrangham and Peterson 1996: 138). Wrangham and Peterson feel that the ordinariness among male orangutans of "raping" female orangutans implies that this behavior is an "evolved adaptation to something in their biology, and this raises the frightening question of whether human rape may also be adaptive" (1996: 132). They then draw the conclusion that humans getting raped should succumb to the will of the rapists.[12]

Feminist evolutionary biologists Catherine M. Drea and Kim Wallen question Wrangham and Peterson's project. They challenge the argument that rape has evolutionary significance. In their discussion of the term *rape*, Drea and Wallen use the term with quotes and reserve its usage for specifically human behavioral contexts, with the caveat that even when limited to humans, the term comes to be a catchall that "subsumes behavior similar in form, but not in function or etiology" (Drea and Wallen 2003: 32). Indeed, the concept of rape has a cultural, social history that entails property rights and the possession of the legal right to consent or not consent (Basu 2011; Bourke 2007; Freedman 2013). Drea and Wallen (2003) carefully examine the evidence of orangutan forced copulation that is marshaled to support the argument that rape has an evolutionary function. They draw an example from John MacKinnon's (1974) orangutan study, which readily identifies an event as "rape" when that event was actually an unsuccessful attempt at copulation. Drea and Wallen (2003) show that MacKinnon's particular example of forced copulation was not a successful reproductive strategy, but instead entailed a male rubbing himself on a female's back. MacKinnon writes, "In one of the observed instances of 'rape' the female continued to struggle throughout and the male's penis could be seen thrusting on her back" (1974: 57).[13] This activity was clearly not reproductive and thus should not be considered an example of a reproductive strategy.

What exactly is forced copulation as defined by orangutan specialists? Primatologist Cheryl Knott points out that characterizing mating events as either forced or cooperative is problematic because copulation between orangutans often entails qualitative characteristics of both (Knott et al. 2010). Knott and her coauthors meticulously categorized characteristics of observed sex acts as displaying four types of behavior: resistance, proceptivity, attraction/genital inspection, and aggression. Their sample entailed

153 encounters observed at Gunung Palung, West Kalimantan, among members of the orangutan subspecies *Pongo pygmaeus wurbii* between 1994 and 2003. Only thirty-nine of the 153 observed encounters were mating events. The researchers further quantified the observed mating events by calculating scores. For instance, they divided the number of resistance acts by the total number of all observed acts to determine how an act would be categorized under their four types. Following Knott, forced copulation is far more complex than the question of whether or not it is reproductive and therefore an evolutionary adaptation.[14] Knott et al.'s (2010) quantitative tools compel me to understand this as a feminist empiricist response to the charge that Wrangham and Palmer (1996) oversimplify. Their oversimplification thus loses empirical precision. Knott's intervention is framed as fact-based evidence over male bias, one that shows that encounters with orangutans are few, and sex acts between them are rare.[15]

The orangutans at Batu Wildlife Center are not confined in a space that facilitates female choice. Rather, as Nadim repeatedly pointed out, free-ranging male and female orangutans encounter one another every day at Batu Wildlife Center. Their conditions more closely resemble the "free access tests" of Nadler and Collins's (1991) study at Yerkes Primate Research Laboratory. Indeed, Nadim's words about female orangutans convey the experience of female orangutans at both a free access test site and at Batu Wildlife Center: "They are free, but in fear." Since reproduction is the sole measurement of success at Batu Wildlife Center, female orangutans' fear and the violence they experience are merely collateral.

Because they are now a critically endangered species, their sexual reproduction is seen by most as a good outcome to be encouraged. The officer Abdullah could thus, in his visitors' orientation, claim with pride that the center is successful precisely because there have been births and there are three generations of rehabilitant orangutans living in the park. It is the same reason Datuk Len could claim that doubling the population of orangutans will be a worthy endeavor for the state. The dangers and violence experienced by female individuals subjected to frequent attempts at forced copulation are not as seriously regarded as the problem of the species' endangered population. This is the context that allows for systematic sexual violence and fear.

The biological anthropologist Barbara J. King (2013) has argued in her National Public Radio blog that while animals may grieve or love, orangutans do not rape because rape is specific to what she calls "unique human cultural systems of power and oppression." Her understanding of rape is based on a philosophical liberalism, one that prioritizes thinking about "free will" at the expense of considering social conditions and constraints. She writes, "[Male apes] may indeed act badly toward females, yet they aren't willfully choosing to inflict harm or violence as an expression of institutionalized male dominance" (2013). King's argument for human exceptionalism is only possible if we ignore the ways in which humans affect the lives of animals, including wild and semi-wild lives.

King fortifies her argument that rape is an exclusively human domain by calling upon humanist feminists, namely a philosopher and a literary critic.[16] The philosopher, Erin Tarver, points out that human language and our notions of gender come to frame and thus inform all such acts of sexual violence. The literary critic Nancy Gray shares the same opinion that language forms our understanding. Both arrive at the same point: when identifying animal sex acts as rape, there is a serious danger that these acts might be seen as natural (and thus inevitable), when it is indeed cultural or social (described through gender ideology and language).

Feminist scholars have long argued that ideas of rape have historical social consequences that cannot be understood through a biological or "natural" framework alone. The popularization of the term *rape culture*, for instance, is indebted to such feminist researchers as Peggy Reeves Sanday (1981) and her cross-cultural comparative work, which showed that the prevalence of rape is not inherently linked to biology and a universal patriarchy, but rather is culturally specific and linked to societal attitudes around violence. This argumentation is fortified by the large body of scholarship on rape as a weapon of war (Card 1996; Feimster 2009; Theidon 2013). How we define rape in respect to legal categories is inextricably tied to histories of property rights, which cannot be isolated from histories of racial and gender oppression.

In the United States, the language and jurisprudence of rape were unavailable to slaves, for instance, who systematically experienced sexual violence but did not have statutory rights to claim victimhood or injury (Davis 1981; D'Emilio and Freedman 1988).[17] The same historical context situates the American cultural concept of southern chivalry, in which white men perpetuate terrorism against black communities for the sake of defending white women's chastity against the possibility of rape (Hall 1983; Wells--

Barnett 1892). I see these examples as a challenge to think about rape through specific and multiple forms of violence.

Instead of focusing on personal experiences of psychological trauma, is it possible to think of rape in terms of forced sexual reproduction and its institutionalization? If so, then female orangutans' experiences at Batu Wildlife Center could be understood as rape, similar to the systematic rape that characterized chattel slavery.[18] Rape was unavailable as a legal term in the context of enslavement, where slaves as property were denied citizenship and the rights of liberalism and were treated as livestock unable to claim legal entitlement to their own bodies.[19] In both contexts, reproduction is exploitation, especially when there is financial interest in having a proliferation of future generations (Baptist 2014; Fielder 2017). In this comparison, female humans and female orangutans bear sexual violence and forced reproduction for the potential benefit of others, not themselves, and with little regard to their experiences of brutality.[20]

Let us pay attention again to Nadim's phrase, "they're forcing themselves to be raped." He uses a passive construction. With it, he identifies victims and survivors, but he does not identify perpetrators and assailants. If orangutans are getting raped, who are the rapists? In this respect, the structural sexual violence of the wildlife center differs from the systematic rape of chattel slavery, in which culpability is obvious: slave masters, overseers, and financiers benefiting from the system of slavery were all responsible. Yet in a context involving nonhuman apes, could there be rape without rapists?

Like Barbara King, I am hesitant to assign intention and say that male orangutans are rapists. Yet my hesitation is not because orangutans do not have will. Rather, I am afraid that identifying subadult male orangutans as rapists ignores the fact that male orangutans are subjected to constraints and violence as well, whether in regard to space or the exercise of behaviors specific to their biological development, or to the contingent combination of the two.[21] In this respect, they cannot be held responsible for the constraints under which they live.[22]

Likewise, I do not think that the people who enforce this structure of sexual violence are culpable for permitting the rape of female orangutans, including prepubescent ones. Blame for this systematic rape should not fall on the people working at these rehabilitation centers alone. Fundamentally, the problem we see here is a much larger problem: the survival of a few members of a species is meant to stand in for the propagation of the entire species.

This ideology is not specific to the wildlife center, but one shared among conservation biologists. These include caretakers of whooping cranes in the United States who withdraw sperm from male cranes and artificially inseminate female cranes for the sake of future generations (van Dooren 2014). Reproduction among semi-wild orangutans forces us to confront the problem that conservation defined through the reproduction of future generations requires sexual violence and what appears to be a gendered distribution of fear.[23]

Violence that is individually experienced but structurally shared by female orangutans calls into question the viability of rehabilitation as a response to species loss. Life at wildlife centers forces the question of how such a life is worth living. Instead of upholding reproduction as the pinnacle of success that comes to justify systemic sexual violence, do we have the imagination to consider other perspectives? Can we imagine other ways of living and dying?

Reconciling Forced Copulation

Workers and managers of the Forestry Corporation carry out protocols that satisfy the sole metric of success, which is reproduction. Female orangutans bear the greatest burden, and the people caring for them know this. Some do so regretfully, others ambivalently, and some with the conviction that sexual reproduction is completely natural, despite the human-directed interventions involved in locking females in with males or keeping them contained in free access conditions within a limited range. How then do foreign volunteers, who are mostly women, reconcile efforts to promote life in the face of extinction by way of systematic rape?

We can begin to answer this question by turning to moments in which they are confronted with questions of life and death, when animals are either free to die or are forced to live and reproduce. I turn to two such moments that I experienced with volunteers: witnessing an occurrence that could have become a sexually violent encounter and experiencing the time a juvenile wildcat was transferred to Lundu Wildlife Center.

Efran, the fully flanged male orangutan, has been in the same enclosure at Lundu Wildlife Center since 1997, when the park began receiving Batu Wildlife Center's excess captive population. Tom, in 2008, thought it would be a good idea to change things up. Everyone went along with his experiment. Through a series of sliding doors linking the night dens to the enclosures, Layang, Ben, and Tom managed to transfer the orangutans to different spaces while preventing access between them. James was transferred to the tower normally inhabited by Ching and Ti, both of whom were seven months pregnant at the time. They occupied his space. Efran gained access to the largest enclosure, normally occupied by the three juvenile orangutans: eight-year-old Lisbet, with four-year-old Miri and three--year-old Gas, all of whom switched places with him. Efran spent nearly thirty minutes exploring the new space: its grassy limits, its different platforms, and finally its gate that connected to the space eighteen-year-old Ching and seventeen-year-old Ti currently occupied.

We—Tom, the volunteers that month, a Malaysian staff member of the volunteer company, and I—stood on the viewing platform and took in the panoptical view it afforded. We could see that Efran was pulling on the iron sliding door to Ching and Ti's enclosure.

Tom explained, "It's enrichment for them [Ching and Ti]. They're probably on the other side thinking, 'Oh my god, he's going to get in!' For Ching, it's probably half fascination and half thought of 'Oh my god, I thought you were dead.'"

The "enrichment" Tom spoke of referenced the common practice in American and Australian zoos in which zookeepers make activities for the captive populations under their care. The idea is to introduce species-appropriate puzzles and other cognitively stimulating objects and encounters in their environments. Such changes to their environments are supposed to help fight the tedium of captivity and curb stereotypy and other psychological pathologies (Braitman 2014; Langford 2017). By expressing what he thought Ching would say if she could, Tom suggested that the orangutans would not know of the other's whereabouts, since they were on opposite ends of the orangutan ape complex. The last time Ching and Efran saw each other face to face was five years prior under captive breeding conditions in which they were caged together without female choice, and which resulted in Ching's first pregnancy at the age of roughly fourteen years.[24]

An ENGAGE staff member pointed out, "He's trying to get to the girls."

Efran sat in front of the barred sliding door. He started to shake it. He began walking around the perimeter.

Tom explained to anyone listening, "He's exploring his new habitat. The idea that orangutans have their specific territory isn't true." At the moment Tom spoke, I didn't know that evidence systematically gathered by primatologists and published that year pointed to the contrary, that orangutans tend to stay within their specific ranges (Singleton et al. 2010; Stumpf et al. 2008). Yet Tom spoke with conviction and authority, even when scholarship indicated otherwise.

A large piece of concrete was loose from the floor leading to the door of the night den. Efran managed to pull it free. It looked like it weighed thirty pounds. It was as big as two human heads. He returned to the door to Ching and Ti's enclosure. Using the full force of his arms, he banged the rock hard and loud against the wrought-iron gate that separated him from the two adult females.

A volunteer who was a British South Asian woman in her thirties and who worked as a medical doctor in England, asked, "What would happen if he got in?"

Only minutes earlier on that platform, Tom had graphically explained to all of us that Efran repeatedly punched Ti on the head during her week in captivity with Efran four years prior. Yet Tom joked, "Make sweet love."

Another volunteer, a British woman in her thirties who had a PhD in zoology, said, "For the girls, not so sweet."

I blurted out, in fear of what looked to be imminent and of what I was afraid to witness: "They're so pregnant. I don't know how that would work. They couldn't be in estrus. They couldn't consent." My blurting out betrayed my own contradictory relationship to endocrinology and the use of hormones to explain sexuality. My sudden response revealed to me and everybody else how deeply ingrained is the language inherited from liberalism on individualistic autonomy and the social contract in which consent is paramount. My blurting out also seemed to suggest that I assumed that sex could only be sexual if procreative. I would never expect such inclinations from humans, nor bonobos. Perhaps I was prejudiced by the literature on orangutans: they were supposed to be different from "GG-rubbing" and "penis fencing" bonobos and humans (de Waal 1995). Pleasurable sex for female orangutans, according to primatological research, is heterosexual sex with the preferred flanged adult male and not with subadult males lacking secondary sexual characteristics (Galdikas 1985; Knott et al. 2010). Efran was the flanged male on the site, but their pregnancies were at such a late stage, their interest in sex was beyond my imagination.

My sudden, publicly expressed conviction wasn't isolated. The zoologist volunteer quickly blurted, "It wouldn't be consensual!"

I began speaking privately with the volunteer, trying to dampen my exasperation: "I don't think this is worth the risk. Why don't they just let them inside? Have they thought about that?"

She responded: "I'm sure they have. They are so concerned about the pregnant ones."

In the meantime, the subcontractor Layang opened the cage for Ching and Ti to escape and enter the night dens. Ti went in quickly as soon as the door was opened. Ching came in once her name was called. Even though they were no longer in the enclosure and were safely inside, Efran still attempted to break the cage door. He banged the rock hard and loud against the iron bars so that it chipped away the iron plate. Later, once Efran was coaxed back into his own den, we found the padlock bracket snapped open by the force of the impact.

Upon seeing the broken gate, Tom joked, "Well, we were trying to test the integrity of the structures!"

Tom felt comfortable joking with us that Efran was attempting to "make sweet love" with Ching and Ti. Yet no one laughed. Instead, I stood in terror at the prospect of witnessing sexual violence committed upon two heavily pregnant female orangutans. Did the two volunteers share my feelings that beholding that sight, participating as spectators to what threatened to be rape, would be terrifying and terrible? Was Tom's joke a way for him to use humor as a way of coping with the conditions of captivity? Or was Tom really so cynical and cruel as to jokingly use such a euphemism? Was his choice of words some kind of glorification of Efran's sexual domination as a flanged "alpha male"? I could not be certain since I was too stunned to formulate let alone pose open-ended questions.

Two days following the incident, I gained a possible answer to the questions I was too afraid to ask. The volunteering zoologist seemed to feel that reproductive success was the most important priority for all ecologically threatened animals, whether orangutans or wildcats. Perhaps for her, what made the possibility of Efran's encounter with Ching and Ti horrifying was not the likely violence of that encounter, but the injury it might do to the well-being of the fetuses in gestation and the impossibility of reproduction that could come of it. The value given to the mothers' well-being lay in their ability to carry those fetuses and others in the future. Reproduction and the future health of the infant, not the well-being of the adult, were seemingly of paramount importance.

The value given to reproduction became apparent to me when the center received a juvenile dying wild cat. It was difficult for all of us to tell whether it was a clouded leopard cat or a marbled cat, both of which were small predator species protected in Sarawak and considered vulnerable according to the IUCN Red List. Sex was in this case easier to identify than species.[25] She was young and very weak with rickets, a calcium and vitamin D deficiency caused by severe malnutrition from having been separated too early from her lactating mother. Staff members and volunteers faced the question of what to do. The feline had no control over the diarrhea seeping out of her. As one volunteer wiped it away, the volunteer with a zoology PhD looked sadly at the feeble kitten and said, "She couldn't be bred, hence it's not worth keeping it alive."

Perhaps through speaking aloud the reference to "she," the volunteer caught herself and corrected herself in the next clause. Perhaps it was too anthropomorphic for her to use "she" when speaking of the feline. Yet the volunteer's statement indicated that the cat's value was solely in its reproductive capacity. Without reproductive capacity, "she," the cat, literally became a worthless "it."

The female body at which we gazed was doubly vulnerable: as a species and as a sickly individual. She could hardly stand, let alone walk. Her body quivered uncontrollably. Tom talked about the incredible pain conveyed by such body language. He suggested caring for it for two weeks, and if there was no improvement, he would resort to euthanasia as an act of mercy. He did not have the chance, though. She died a few days later. The hope for self-recovery paired with the fear of killing her made it possible to prolong the cat's pain. That to everyone was better than facing the responsibility of having killed a member of a threatened species.

Conclusion

Reproductive capacity is the highest priority for wildlife conservation in Sarawak among Forestry Corporation officers and foreign volunteers. Heterosexual sex is seen as inherently natural, despite the human interventions that shape the context in which reproduction can happen, whether that is forcibly holding a female and male captive together in a cage, facilitating sex with a prepubescent female orangutan, or holding multiple orangutans in free access conditions better suited for a single orangutan. Sexual violence experienced by female orangutans comes to be a "natural feature,'" despite the specificity of the situation among semi-wild orangutans for whom forced copulation is more typical than in wild conditions. The brutality perceived in these acts of copulation is seen as either an instinctual act of nature or as a necessary collateral outcome for the purpose of conservation. Forced sexual reproduction highlights the underlying violence that occurs when the response to the threat of extinction is to increase the population of an endangered or threatened species. Artificial insemination, although less obviously brutal, is nevertheless still violent (Blanchette 2015).[26]

When Nadim in our conversation stressed the death of the ecosystem, as evidenced by the profusion of pregnancies and the sense of fear and hatred he perceived emanating from the faces of female orangutans on site, he offered a way to think systemically that I think we need to take up as an alternative response to the threat of extinction. If we recognize the consequences of burdening individuals with the responsibility of bearing the future of their entire species, if we consider how more births could lead to a dead ecosystem, we could then be inspired to redefine and expand what it would be to live and flourish in the face of impending deaths.

Chapter 4

FINDING A LIVING

Lundu Wildlife Center is both a site of refuge for animals that have lost their habitats and a workplace for Sarawakian Ibans who felt pushed to relocate from their ancestral homelands in the interior, who were unable to maintain a livelihood through subsistence farming, and who must earn the means to live in Sarawak's cash economy. Sarawakian animals and Sarawakian people gain new subjectivities through displacement. The Malay idiom *cari makan*, which literally means "to find food," is a common metaphor in Sarawak and the region for earning wages.[1] The idiom suggests that something is gained through the loss experienced by both displaced animals and the people who are paid to care for them. Iban workers at Lundu forge new kinds of social relations. These relations are new in that they interact with animals in ways that radically depart from their own historical, folkloric traditions of connecting across species. What might have been a reverence for orangutans as well as crocodiles, sun bears, and omen birds in the past appears to fall away amid everyday confrontations. Through ecological loss, displaced animals and displaced people gain a shared interface, which in turn produces new ways of becoming. In other words, loss produces new environmental subjectivities. In this chapter, we explore how colonialism continues to structure life and death in officially decolonized Sarawak.

The possibilities opened through an interface of loss confronted me one afternoon at Lundu Wildlife, in the waning wet season of February 2010. I accompanied workers drawing blood from the crocodile named Bako. Bako was named after the location in which "it" was found.[2] A fisherman off Sarawak's coast pulled in his net and accidentally caught the six-foot--long false gharial, an endangered crocodilian species under protection in Sarawak. As a protected species, hunting, killing, selling, or eating it could result in a yearlong prison sentence and a ten thousand RM fine, which was almost US$3,000. The fisherman called the Forestry Corporation and had them take it away. Thus Bako was sent to Lundu Wildlife Center, where it was to live in a fenced-in enclosure with three other crocodiles. The enclosure was built prior to 1997 on cleared jungle soil; concrete had been poured over the exposed forest ground in order to accommodate the crocs' pool.

At the moment I was looking at Bako, it—for it was impossible for any of us to tell its sex on sight alone—was hog-tied and waiting to have blood drawn from a spot on its back just below its skull. The blood sampling was part of a joint study conducted between the Forestry Corporation and the University of Malaysia Sarawak, although neither institution technically employed those who were collecting the samples. That job of preparing the crocodiles mostly fell upon the corporation's Iban subcontractors stationed full-time at Lundu Wildlife Center: Layang, Ngalih and Ren. Ben, the Bidayuh program facilitator of the British volunteer company, ENGAGE, helped the three men with their work. They all followed the lead of Layang, whom they mockingly addressed as *pakar*, the expert. One man would lasso the croc's mouth shut; the crocodile would arch its back and neck upward. Someone would then toss a burlap sack over the crocodile's eyes. On the count of three, they would jump in unison onto the croc's body. One man would begin tying up the crocodile, starting with its mouth, and then would pass the rope onward. At one point, a rope snapped and everyone scrambled off the crocodile—Layang was the last to jump off as it scurried away to the muddy pond bank of its enclosure. Bako was the first crocodilian to be hog-tied; its heavy hiss-like breathing punctuated the time until the veterinarian from Kuching came to take the blood sample.

Cindy and I waited near the fence of the enclosure as the men wrestled with the other two false gharials. She was my peer in that we were born the same year, and I was mistaken as sharing her ethnic Chinese-Iban background. She was also my informant, particularly in her capacity as a junior officer of the Forestry Corporation. We stood far enough from potential danger had Bako or any of its enclosure companions broke free from their thin rope bonds. As Cindy stood beside me, she explained to me how Bako came to live here. Ending the story about the fisherman's mistaken catch, she said, "Why did he just not throw it back? It's good they called us, but it would have just been better for everyone." The "everyone" to whom she alluded included the crocodile.

At the end of the day, well past quitting time, the blood sample was too difficult to get. The crocodiles were released after an educational exercise for the keepers involving Bako's terminally ill enclosure mate, Stampin. Since that crocodile was thought to be at the threshold of death, the veterinarian figured its demise could be a pedagogical moment for everybody present. The vet wanted to show the crocodile's circulation system, its nervous system, and how one could paralyze a crocodile with a syringe by puncturing the spinal cord. Doing so demonstrated the expert care necessary for an attempt at drawing blood from a cold-blooded animal, and it was expertise that he felt sufficiently authorized and inclined to impart. Stampin kept moving long after its spine was supposedly punctured. Fearing that this act of euthanasia had not worked, and with Tom expressing concern about the ethics of what might be a live animal capable of feeling pain, the veterinarian ended up using a hammer and chisel to break the spine. Ben would later tell me how every strike of metal upon metal made him secretly wince, thinking about the nape of his own neck.

These encounters open difficult questions: How does one determine who has the right to kill or let live? This was a question faced by the fisherman, who was afraid of a prison sentence, and by the veterinarian administering euthanasia—a death intended to be ethical. How do empathy and identification grow in this interface, as embodied in Ben's response to the clinking of chisel striking bone? How does the performance of gender shape orientations toward animal others, as Cindy and I demonstrated when we kept our distance from the men wrangling crocodiles? How does a failed lesson shared by an urban veterinarian with rural workers without formal education simultaneously perform and undermine expertise? These questions can begin to be addressed only if we address how the relations taking place here can even occur. This requires an inquiry into the forms of loss that produce the interface of these different confrontations and encounters.

Loss for Iban workers at Lundu Wildlife Center generates new relations, and with them new formulations of the self: first as environmental subjects who must cari makan, or find food in ways that differ from the past, and second as wage-earning animal handlers who must interface directly with animals that they would otherwise rarely encounter. This interface gained through environmental loss is the source of expert, experiential knowledge. Wrangling crocodiles, as Layang and his coworkers in the crocodile enclosure shows us, is a manifestation of expertise, of being pakar. Likewise, the attempt to paralyze a crocodile using a needle and syringe was an act of expertise, regardless of its apparent lack of success.

People and animals at Lundu are both environmental subjects, a term taken from Arun Agrawal (2005), who offers it to identify new ways of becoming by way of relations with governed environments. Influenced by Michel Foucault's (2007) ideas of subjectivity and governmentality, Agrawal sees forest regulation as subject formation. Whereas Agrawal focuses on the transformation of people through their changed relationships with forests in northern India, my intention here is to think about the wildlife center as an institution that shapes the lives of both human and nonhuman subjects that move within and beyond its confines.

In other chapters, we see how certain subjects are able to move beyond the site: commercial volunteers who stay for two to four weeks, paying a thousand dollars a week for their stay; the expatriate British manager of ENGAGE, Tom, whose paltry wages supply him with his simple needs for coffee and inexpensive cigarettes smuggled from Indonesia; urban-based and college-educated officers like Cindy; and myself, a local-looking but foreign ethnographer. My focus in this chapter is on those who remain within or at the park's boundaries: the animals on site that have nowhere else to go, and the economic migrants to urban outskirts whose wages compel them to stay. Animal and human environmental subjectivities meet here because of their shared histories of loss.

This chapter traces how people new to wage labor and animals formerly wild and newly confined must now cari makan, or earn a living, in the same site. In what follows, I show how the idiom of cari makan, as expressed through encounters with illipe nuts, turtle eggs, and crocodiles produces environmental subjectivity through displacement. I then examine how the work of animal husbandry involving orangutans and sun bears produces knowledge or expertise: what may appear as domination is indeed unequal risk involving mutual vulnerability. Here, expert knowledge obtained through new relationships is gained through displacement. Ultimately, I show how relations that create subjectivities at Lundu Wildlife Center are produced through ecological loss and new contexts of wage labor.

The understanding of loss as generative has an affinity with David Eng and David Kazanjian's (2003) *Loss: Politics of Mourning*, in which the authors see that absence is always potential presence. Loss for them captures the senses, discourses, and processes lingering in the terms *mourning*, *melancholia*, *nostalgia*, *sadness*, *trauma*, and *depression*. The question driving their inquiry is not about *what* is lost, but rather *how* feeling loss and naming that loss produces what remains: ambiguous affect in which mourning, melancholia, and militancy can blur together and offer political possibility.

Following Eng and Kazanjian (2003), my attempt to highlight loss as a shared experience across species should not be mistaken for an attempt to salvage what once was or what it had aspired to become. This is not a eulogy about old ways of life in the longhouse (Sahlins 1999). Nor is this an example of "imperialist nostalgia" for lost native culture (Rosaldo 1989).[3] Nor do the threats of extinction and "anticipatory nostalgia" around potential extinction loom in the lives of Ibans as it does for orangutans (Choy 2011; Sodikoff 2012).[4] Instead, my goal is to highlight what is shared between endangered animals enclosed within Lundu Wildlife Center and the people who care for them. I am interested in how new ways of living together are produced through histories of violence—slow and uneventful, as well as eventful and dramatic. Violence appears through the new economic and spatial contexts in which one must find food and through the direct engagement of animal husbandry, which is the source of wages for those in this new context.

Loss through displacement is the exemplar of "slow violence" for literary critic Rob Nixon, who was inspired by anthropologist Adriana Petryna's ethnographic subjects, Chernobyl victims claiming and being denied biological citizenship (Nixon 2011; Petryna 2002). The term calls attention to how violence works over multiple time scales and beyond the clean boundaries of specific events, places, and bodies affected. For instance, the era of cable news shapes our attention toward events for which our memories can be recalled with handles and hashtags, whether 9/11, avian flu pandemic, or the BP oil spill. Invoking slow violence addresses the challenge in representing uneventful but nevertheless devastating crises, such as lingering radioactivity or the loss of species and with it biodiversity due to the destruction of their habitats for agro-economic interests (Hong 1987; Masco 2006; Oliver-Smith 2009; Petryna 2002). Nixon points out that displacement, in all its temporal, geographical, technological, and rhetorical forms, is materially experienced through eviction, mega-dam construction, and natural resource extraction. Instead of thinking of displacement as forced movement, Nixon sees it as "the loss of the land and resources beneath them [those who must move], a loss that leaves communities stranded in a place stripped of the very characteristics that made it habitable" (Nixon 2011: 19). This understanding of displacement forces us to think about the particular devastation and slow violence caused by material, environmental displacement.

A destroyed forest converted to a dam or plantation cannot resume its previous form within the lifetimes of those who experienced the loss of those forests, despite the agency plant life, wildlife, and other life forms exhibit when thriving and taking over crumbling structures and facades in abandoned man-made sites (Stoler 2013; Vergara 1999).[5] What remains instead is an environmental subjectivity produced through a history of displacement, and with it transformation.[6]

Cari Makan: Finding Food, Earning a Living

Lundu Wildlife Center was founded when Batu Wildlife Center exceeded its carrying capacity for its free-ranging orangutan population. Orangutan rehabilitation efforts and the responsibility for holding protected species were transferred to the new center in 1994. The intention at its founding was for Lundu Wildlife to eventually host a population of free-ranging orangutans, alongside its population of animals exhibited in enclosures: crocodiles, pigtail macaques, pangolins, binturongs, and other members of protected and totally protected species.

At Batu Wildlife Center, a routine exists that is known both by the orangutans and their keepers: visitors start to come at 8 AM for the 9 AM feeding. A couple of the younger orangutans are there first. The two orangutans who have been there the longest, Deh and Grandma, will make their way together from the north, above the covered bridge or on it, and then to the central courtyard, where they'll go to the shady tree or up to the ropes above a small feeding platform. All that time, visitors will take snapshots, laugh at their motions or expressions, tell each other narratives of what the animals are thinking, and ask the workers questions. The Batu Wildlife Center workers answer while at the same time trying to maintain distance between the animals and the visitors. They will instruct the visitors where to stand in relation to the orangutans. The orangutans may move in one direction or another and may just decide to sit still on a pathway. Then it all happens again in the afternoon from 3 PM to 4 PM.

This is the kind of routine in which everyone regularly on site, human and animal, has to cari makan.[7] In Malay, cari makan literally means to find food, and figuratively it means to earn wages. The orangutans know that a consistent food source is at the courtyard and feeding platform where anywhere from a handful to a hundred visitors, many of whom breach the five-meter separation they are supposed to leave between themselves and the orangutans in order to take photos and observe them. The workers know that their job is to feed the different orangutans while trying to keep the visitors safe. Ricky once joked that the orangutans' job is to appear, eat, and pose for the camera. At Lundu Wildlife Center, the animals' jobs are to sit, eat, and wait. For the most part, the furry, feathered, spiky, and scaly residents at Lundu Wildlife have nowhere to go but their cages and enclosures.

Apai Julai has worked at Lundu Wildlife Center for a wage ever since construction began in 1993. He has also been the one and only headman or *tuai rumah* of the Iban longhouse just outside the wildlife center's boundaries. Indeed, his role as a wildlife center worker was due solely to his status as the closest community's recognized leader. Syed, the person in charge (PIC) of Lundu Wildlife, himself an officer trained in the Forestry Department and later the privatized branch of the department known as the Forestry Corporation, felt it was important to enlist community involvement and local support for Lundu by having the center extend material benefits through paid employment. Apai Julai fibbed about his age to keep working at the center. Officially he was fifty when he was really fifty-eight at the time I met him in 2008. When the Forestry Corporation's shrinking budget could no longer include wages for Apai Julai, despite the paltry sum of about 800 RM a month, Lundu Wildlife's PIC asked for ENGAGE to continue employing him.

Layang's wife Mit once joked with me that people respect the Tuai Rumah Apai Julai, but animals do not.[8] In the span of two years, Apai Julai had been bitten three times, each time by a different animal. The first time was a bite from a gibbon, the second time was from a macaque, and the last was from Ching the orangutan.[9] He suffered through these bites with resilience. He was a tough, gruff man and bites were part of his employment for wages.

Apai Julai, his brothers, and his parents came from Batang Ai and were the first to arrive in 1977 at what became the Iban longhouse settlement Kampung Mohon.[10] Relocating entailed taking a boat, all of one's worldly possessions, and enough rice to last until more could be planted. Explaining the motivation to leave, he said to me, "at Batang Ai, we were *ulu* [far upriver]. Here we're near the *bandar* [city]. Over there, it was hard to cari makan."[11] They could no longer "find food," as he explained to me using the Malay idiom for finding a means of living. They were unable to make a living, literally and figuratively.

Shortly after Apai Julai's family arrived, other Ibans from other upriver interior regions joined them on the outskirts of the city: Ulu Sebuyau, Betong, Sri Aman, Lubok Antu, and Batang Ai. Their dialects of Iban were distinct from one another and differed in lexicon. Yet they all came for the same reason: they could no longer cari makan or find food in the forests of their ancestral homelands.

Social scientists studying Iban communities in the 1970s reported severe rice shortages. The Sarawak Museum fact-finding team charged with the task of assessing the potential impact of the then-pending Batang Ai Dam noted that very few households in 1977 reported a sufficient rice surplus to be able to barter or sell (King 1986; Sarawak Museum 1979). R. A. Cramb (1979), in his report to Sarawak's Department of Agriculture, where he worked as an Australian Volunteer Abroad between 1977 and 1980, noted that only 20 percent of households in the Batang Ai area fulfilled their needs for rice. Most households met a mere third of their rice needs and either supplemented their diet with less favored foods of manioc, taro, and sago, or they bought rice by engaging in wage labor and selling cash crops like pepper. Cramb (1979) reported that the average yield of rice in the 1976–77 season was fifty *gantang* or gallons per acre. The lost yield appears especially devastating when compared to Derek Freeman's report of an "unusually poor season" in 1949–50 of 118 gantang per acre (Freeman 1970).

Apai Julai's explanation that it was hard to find food might have been all too literal, although the figurative sense of cari makan was also significant enough to warrant relocation near the city. Kampung Mohon's *van sapu* driver Rangkai explained this to me in no uncertain terms as we drove on his route picking up passengers from Kuching to the village in which we lived, dropping off passengers at other settlements along the way.[12] Conversing with me as I sat in the passenger seat beside him, Rangkai emphatically waved his arm toward the hilly forests in the horizon. "*Begaimana cari duit di hutan*?" (How to find money in the forest?) He came from the swampy peat forests of Ulu Sebuyau as a boy in the 1980s, but what he had to say pertained to everyone in Kampung Mohon.

My friend Rangkai the van driver did not see the forest as a source of wealth, even though his having gone on *bejalai* to Papua in his youth in the late 1980s and early 1990s, where he worked as a lumberjack at a timber camp, enabled him to achieve his present circumstances.[13] He saved enough to buy a truck to transport villagers to and from their rented rice paddies an hour east of the village, for which he charged a fee. He then bought a rice huller machine for which he charges his fellow villagers. The revenues from that went to the purchase of the van in which we sat, and eventually a *kedai* where his wife sells dry goods and a few refrigerated drinks. His parents, who live with him, grow rice for their shared household's needs. Like nearly everyone else at Kampung Mohon, they do not buy rice but instead grow their own on rented land. Unlike most at Kampung Mohon, he is able to send his daughter to a technical school in Kuching, where she studies to become a nurse. For him and others, proximity to the city, to the bandar and the marketplace, is the source of money in its hard cash form: *duit*.

Within Apai Julai's and Rangkai's lifetimes, Sarawak has witnessed profound transformations, facilitated by the arrival of motorized boats, chainsaws enabling timber extraction, and overland paved roads leading to the quick selling of massive volumes of raw goods to marketplaces (Kaur 1998). This is not to say that the upriver interior people of Sarawak lived in isolation prior to their relocation to the bandar. Indeed, they have long been connected to global markets, as evidenced by the trade of jungle produce for heirloom pottery from China dating to at least the 1400s (Dove 2011; Padoch and Peluso 1996). Throughout Sarawak's history as an autocratically governed Raj from the 1840s until World War II, smallhold producers have brought pepper, rubber, and other cultivars to international markets (Baring-Gould and Bampfylde 1909; Freeman 1970; Grijpstra 1976; King

1993). Paired with this history is the Iban gendered cultural practice of *bejalai*, the practice in which men migrate for prestige, by means of which men like Rangkai in their youth were able to amass personal wealth. Such a cultural practice among men has a long history over multiple generations. Yet faster and larger-scaled forms of extraction and development marked the second half of the twentieth century, following Sarawak's incorporation into the Malaysian nation-state, than had previously been experienced. This is especially apparent in development projects involving dam construction.

Between 1982 and 1985, for instance, Batang Ai Dam displaced 3,000 Iban people, including some of the first residents of the longhouse just outside of Lundu Wildlife's gates, of which Apai Julai is the headman, himself having been displaced from Batang Ai. Batang Ai Dam flooded 24 km^2 of primary forest inhabited by orangutans and other fauna, and affected a catchment of 1,200 km^2. Resettled people were promised 17,000 acres of land, but only 7,600 acres were feasible for resettlement (King 1986: 15). Along with the Sarawak Land Consolidation and Rehabilitation Authority's push toward participation in the agricultural economy by bringing goods to market, land shortages coincided with agrarian labor shortages associated with bejalai.

Batang Ai Dam not only displaced people, their ecologies, and their economies; its creation directly displaced two orangutans at Lundu Wildlife, Ching and Ti. A luxury hotel built near the dam allegedly found the two orphans and surrendered them together in the 1990s, after an unknown period during which the orangutans earned their keep as tourist attractions at the hotel. It goes without saying that many more animals and other life forms likely met their demise. Despite the destruction experienced by resettled people and displaced animals, Batang Ai served as a model for Bakun, an even larger-scale dam built three decades later.[14]

In the years since Apai Julai's resettlement with his family in Kampung Mohon, the two-day riverine connection from their new home to the capital city of Kuching would be replaced with a two-hour drive in the 1980s. A national park would be established on the other side of the River Mohon by 1989. A domain on the park closest to Kampung Mohon would become an orangutan rehabilitation center by 1997. Apai Julai could declare to me, with exactitude about year and even date, when he founded the village: April 28, 1977.[15] In the same way, he could pinpoint the year in which they "entered" (*masuk*) Christianity through Anglican proselytizing by Iban missionaries: 1980.

The crisis of food shortages and the inability to sufficiently participate in a cash economy while upriver coincided with the political crisis of communist insurgency and the Cold War. Apai Julai explained to me that they were encouraged by the government of Sarawak to relocate because of the communist insurgency that began in 1962, in the period of transition from British colonial rule to the rule of the federal state of Malaysia.[16] With war being waged in the forests on the border between Sarawak and Indonesia, orangutans, too, felt the consequences of the Cold War: an infant orangutan served as a living mascot to the Royal Army's Troop C at Batu Lintang, which was only a few miles from the Sarawak Museum in the city center of Kuching. When the orangutan infant was surrendered to the museum, it was bald, constipated, and underfed (Sarawak Museum Curator 1965).[17]

Violence continued into the 1970s, despite the fall of the leftist Sukarno in Indonesia and the ascendancy of Suharto's CIA-backed military dictatorship that formally ended the confrontation between Indonesia and Malaysia. The 1970s witnessed a series of state-supported military operations against communists in Sarawak. It was estimated that five hundred communists operated in the state, working on both sides of the border shared with Indonesia. Five thousand Malaysian troops were deployed for counterinsurgency efforts. Some operations, such as Operation Jala Raya (Malay royal net), increased policing by placing curfews, conducting house searches, patrolling jungles, and bombing communist hideouts. Others, like Operation Ngayau (Iban: to hunt or go on a war party), attempted to undermine communist support by promoting rural development. Rural communities received treatment from medics while army engineers built roads and bridges for local use. And other actions, like Operation Pumpong (Iban: to decapitate), forced the resettlement of specific longhouses.[18]

Living in a new region, one neither fully urban nor upriver, between and yet remote from both Lundu's town center and the city of Kuching, entailed finding new forms of occupation. The location of Kampung Mohon afforded the possibility of maintaining some aspects of their former lives. In the forests nearby, they could collect *buah kabang* and other produce for both their own consumption and commercial trade. In the season for buah kabang or illipe nut, old people would spend entire days collecting them or drying them in the sun. Children coming home from school would peel the dried casings well into the night.

Buah kabang is a cultivar thriving in the mixed and anthropogenic forests of Lundu Wildlife Center, planted by Kampung Mohon residents before the boundaries of the park were established. Even after the forest near them was officially made into a "totally protected area" as a national park, Apai Julai said they were still able to obtain such items for personal and commercial use, though they had to hide it from park authorities. Coincidentally, an incident occurred while drawing blood samples from the crocodiles. The efforts extended beyond quitting time and Ren was not particularly keen on watching a syringe being repeatedly jabbed into a roped-up crocodile. So Ren started collecting ripe buah kabang from the ground. When Lin, their boss from the Forestry Corporation, saw this, she asked why he was collecting it. He said it was to feed the sambar deer, which the park underfed because they had to stretch the food budget and for which the ENGAGE volunteers collected their food scraps to feed. Had he not collected it, it would just fall to the bottom of the crocodiles' cement pool and clog the drainage, which he would eventually have to clear himself. Collecting it, he was able to put food on the table, literally since it can be sautéed, and figuratively since it can be sold once shelled and added to the bushels picked by his family.

In the context of having to conceal his commercial use of the forest, however small in scale, Ren and the residents of his longhouse have to find multiple means to cari makan. By the time I arrived in the longhouse and began living as an adopted daughter with Ren, his wife 'Kak, and their family, the longhouse's sources of income included remittances from daughters working in Kuching's plastics factory or its fast food eateries, as well as from sons working in Papua for logging companies, or in peninsular Malaysia as soldiers or industrial workers. But for the most part, cari makan meant working in construction for the many building projects that sprouted up on the outskirts of Kuching since their arrival in the region in the early 1980s or at the logging concession operating nearby at the other corner of the national park. For a few, like Ren and his brother, it meant working for the Forestry Department's Parks and Biodiversity Conservation Division.

Before orangutans, caring for sea turtles paid some people a wage from the Forestry Department. We spoke about it when Kelvin Egay, my Kelabit friend who is also an ethnographer, paid a visit.[19] As we drank *langkau* (rice whiskey) together with my *kaban* (kin), Kelvin started asking questions about turtle eggs, including their current cost. Turtle eggs are a local delicacy, but illegal to consume; their consumption is seen as the reason for the turtles' declined numbers. Long gone are the days when everyone, including the conservationist Tom Harrisson, could participate in turtle egg throwing and eating festivities at the start of the turtles' nesting season (Heimann 1998). Ren's older brother said, "We don't like talking about this. Not even here in the longhouse, among ourselves. People will get angry if they know we are talking about this. We can't talk about turtles, orangutans, and hornbills." My friend, in his jovial way, managed to push the conversation further about turtles and their eggs. Ren said that he doesn't like the consistency of turtle eggs because "it's like snot." Kelvin then said, "But does it taste good?" To which Ren replied with laughter, "Yes!" Later my friend shared his analysis with me: "They're all conditioned by conservation so that it's now in their *adat* [customary law] to not talk about it."[20]

The self-imposed taboo against speaking about totally protected species at first appears to be an example of how individuals in a community have become what Arun Agrawal calls "environmental subjects." Yet Kelvin's critique seemed to contradict Agrawal's analysis by pointing to the difficulty of ascertaining one's subjectivity via discourse alone. Kelvin's remark points to what appears to be a mere adoption of environmental discourse by Ren and his kin. This raises a serious question: to what extent do individuals like Ren become environmental subjects? Ren could still attest to the pleasure of eating a turtle egg, despite its viscosity, its illegality, and its implications for the turtles' conservation.

Ren's experience calls for expanding, not limiting, our ideas of environmental subjectivity. Ren is not an environmental subject just because his source of income is tied to conservation associated with flagship species. Rather, he is an environmental subject because his subjectivity is constituted through the slow violence of ecological loss. Ren's laughter shows the complexity of what it means to feel the impact of environmental damage. In his lifetime, Ren has experienced an utterly transformed relationship with ecology. The reduced viability of living the "old way"—of hunting, tending *padi* [rice fields], cultivating the *kebun* [garden], and foraging for bee nests and other forest products—forced his family to move in his youth. And now he must use environmental discourse to defend his foraging activities in the forest; his foraging can no longer serve his family's needs, but must be justified according to the needs of the animals he is paid to tend. Conservation is just one "new way" of earning a living, alongside construction work and logging. The work of conservation and animal management is not his source of identity, pleasure, fulfillment, or other forms of his subjectivity. His laughter about enjoying the taste of fertile turtle eggs even while he is employed to care for their future has to be understood in the larger context of mass displacement and relocation. The egg-laying turtle, the crocodile, and the caretaker are all contingently brought together, in entirely different contexts from whence they came, in which ever returning to what was lost feels impossible, and it remains unclear as to what is to become.

Expertise

It was my adoptive younger sister's birthday party in Kampung Mohon, and Lisa, our cousin in her late teens who lived next door, helped us cook and clean up. She had worked at an international fried chicken chain in Kuching. She had lived above the restaurant in a dormitory with other young female coworkers before they all quit at about the same time, which was shortly before the rice harvest festival Gawai. While we worked together in the kitchen, she showed us how to properly wash dishes—despite our daily participation in household chores. First, she boiled water in the electric kettle. She then used it to fill a flat plastic bin with hot soapy water. She dropped all of the dirty cutlery in it, shook it vigorously, and dropped it in a clean bin of water. It was very efficient. Another cousin commented: "*Pandai sebab* KFC"—"she's an expert because of KFC." The emphasis in the sentence was not the word *pandai*, meaning clever, expert, and skilled.[21] Rather, it was KFC, said slowly with a lilt, signifying global cosmopolitan taste. Lisa's motions and instructive lessons demonstrated how wage labor made her an expert. The unpaid work we performed at home did not.[22]

Expertise here is authority acquired through experience, not formal education, and is tied to wage-earning work, as much for the young women in Kampung Mohon as it was for those earning wages at the wildlife center a stone's throw away, including Layang. Expertise is hard won through experience, yet it is not sufficiently recognized through prestige and monetary compensation. It is recognized in name only, and being named an expert is a form of derision and mockery. Thus, my cousin could smirk when she said Lisa's expertise came from KFC, just as Ren and the others called Layang an expert when he took the lead in wrestling crocodiles.

The push to participate in wage labor at Kampung Mohon is not quite an example of proletarianization, well documented in the earlier works of anthropologists publishing in the 1980s (Comaroff 1985; Ong 1987; Taussig 1980). My cousin Lisa was not disciplined into becoming a docile worker in a consistent workforce. After all, she quit her job just in time to get her paycheck before Gawai. Although Layang and my adoptive parents worked for wages, the labor in which they were engaged was not the mechanistic Fordism of the assembly line. Far from the monotony of the factory or the service work of mechanized fast food, the work at Lundu Wildlife Center entailed dangerous forms of uncertainty in which experience leads to expertise. Workers become knowledgeable about animals while animals become knowledgeable about the workers who look out for them. It is here, at this dynamic interface, that we can see what environmental loss generates: the push to become pandai, in which animal and animal handler become experienced with the other.

Layang was seventeen years old when he began working in orangutan rehabilitation. On his first day of work, without any training, he was given a broom and dustpan and told to shovel scat in a cage in which eight fully grown adult orangutans were kept. The immersion through which he learned how to become an animal keeper was difficult to demand of anyone. He was one of only two at Lundu Wildlife Center who had the experience that could facilitate the courage to enter a cage with an adult orangutan in it. It was for this reason that I was encouraged to follow Layang in his work activities after I began field research in October 2008. Every day that I spent with him, I saw how he exercised his expertise and courage with other animals on site.

Lundu Wildlife Center had other animals in states of captivity and not enough space to adequately house them. For two months, a sun bear was kept in a cage within the orangutan night den. The bear simply had nowhere else to go. He was confiscated from an illegal menagerie, and the other two bear enclosures each had four bears in them. Sun bears of Borneo may be the smallest bears of the world, but they are reputed to be the most aggressive and territorial of all bears. Their vicious-sounding barks carried through the air. Sun bears and orangutans would rarely interact with each other in the wild, since orangutans are active in the day while sun bears are usually active at night. While the orangutans at the rehabilitation site were able to go outside to an open-air walled enclosure, the sun bear in the orangutan night den was kept in his cell twenty-four hours a day, seven days a week. His cage had a two-square-foot barred door that looked onto Lisbet the orangutan's enclosure. Whenever Lisbet saw the bear, she would take a rock and throw it at the enclosure from a distance; the bear would retreat deeper in the cage, and Lisbet would scurry away.

On my first day observing animal husbandry, Layang entered the bear's cage when the bear was still in it. This was very shocking to me at the time because I had just barely ended my zookeeper internship at the Oakland Zoo in the United States, where I cared for a sun bear. As at every accredited zoo in the United States, encounters with animals are mediated through sliding doors and mesh fencing. Entering a cage with a bear in it is akin to risking one's own life.

Why did Layang risk his life by entering the bear's cell? The answer might have been his interest in demonstrating his own expertise, to prove his fearlessness at having physical proximity to an infamously aggressive creature. Perhaps this demonstration of fearlessness was provokingly gendered and classed in opposition to my tech-wear clad, bespectacled, American--educated, but local-looking self. Undeniably, there was a simple answer: it was to clean its dwelling. This was a major part of the job of working as an animal handler at Lundu Wildlife Center.

The cage was as small as a prison cell: no more than 10' × 10' × 10'. In one hand Layang carried the small fire hose, and the other hand held half of a small watermelon. He opened the cage and presented the watermelon to the charging bear. With one hand, Layang pushed the watermelon toward the bear's face, projecting the watermelon away from him. He then used the hose to wash the floor and the two metal platforms attached to the wall inside the cell. Whenever the bear stepped near him, he doused the bear's face with water. Once the bear finished the watermelon, rind included, he went for Layang, who would spray him directly with the hose. It looked dangerous enough for me to offer Layang sugar cane from the bucket of food I carried. He declined, using the hose as his sole defense against the approaching bear. A few seconds later, Layang gestured a request for the sugar cane. I gave it to him and he placed it on the top platform. The bear went for it, but his paws couldn't get a proper grip on the wet and slippery platform, so he fell and hit his head on a metal beam. Layang laughed while moving the sugar cane from the top level to the level on which the bear writhed. The bear tore into it. He then closed the gate, which had been open the whole time, and secured the lock.

It was difficult for me to understand. When Tom asked how accompanying Layang had gone, I shared my surprise at Layang entering the bear's cell with the bear still in it. He said, "The other option is to have the bear live in his own shit." Contrary to Tom's opinion, washing bear feces could be done *without* risk of injury by taking the hose and carefully aiming and shooting the droppings away. Layang, however, seemed to want to demonstrate the more thrilling option.

Those familiar with Iban ethnology and ethnography might note that the properties of Layang's expertise appear to be part of a long-standing performance of his gender as an Iban man. Writing about the practice of young Iban men felling giant trees between 1949 and 1951, Freeman writes, "Felling such trees, often on steep and broken hillsides, is an undertaking which calls for strength, dexterity, agility, and nerve. But these are qualities which the Iban cherish, and which every young man of ambition hopes to develop" (1970: 173). Fear was never something owned by Layang. Instead, he seemed to espouse the opposite with his sense of humor. Every morning upon arrival at the quarantine kitchen, he would join his peers in preparing the animals' breakfast, and he would announce his daily Radio Sarawak impersonation: "*Selamat Pagi, Bumi Kenyalang*" (Good morning, Land of the Hornbill), which would always make everyone laugh.

The aspirational characteristics of strength, dexterity, agility, and nerve were achievable in the work Layang performed as an animal handler at Lundu and Batu Wildlife Centers. These were the characteristics of his becoming pandai at the job of animal handler. Likewise, the young sun bear, in having his space encroached upon by Layang, was forced into a position of having to tolerate Layang's intimate co-presence. The bear, too, had to become pandai with humans. Being co-present, doing one's best to anticipate the next move, are all part of gaining the expertise of experience. This particular expertise is about being radically vulnerable to the other, when the threat of injury is real. It is not an equal form of vulnerability. It is Layang who laughs at the bear slipping and falling—not the other way around. And yet Layang is not safely in a position of domination. The only barrier between himself and the bear is half a watermelon rind and a fire hose that has as much water pressure as a garden hose. This is shared but unequal vulnerability.[23]

Layang's performance of expertise was exhibited through his willingness to experience vulnerability, which he contrasted with the performance of expertise through credentials. On the first day I met him, he had just come back from a week of attempting to train the orangutan Lisbet with the Kuching-based conservation director. The conservation director had an MSc in biology and rarely left the office for fieldwork. The contradiction between official credentials and experience was a point Layang loved to mock. He laughed at how the conservation director tried to command the then eight-year-old female orangutan Lisbet to climb up a tree by pointing up above his head and yelling at her. Layang especially loved the part of the story where Lisbet responded by slapping the conservation director on the face.

Most of the time, Layang's experiences with hierarchy and expertise were far from funny. In the wet season of 2008–9, Layang and Ben from ENGAGE wanted a new ranger station deeper in the park, further away from the enclosures visited by people. There they would provide food for the orangutans twice a day. But the Forestry Corporation did not want to release them because they were afraid of liability should visitors be injured by a free-range orangutan. Years ago, when the Forestry Department became partially privatized, Ching and Ti had to be released to the park because there was not enough food to support a captive orangutan population. Ching was pregnant then and successfully had Mut while she was bebas and lacked food support. Not having enough wild food sources or perhaps having little interest in foraging, Ching would often stay around the wildlife center. She developed a reputation for attacking visitors.[24] The corporation was not interested in Layang's opinion, because his eighteen years of experience working with semi-wild orangutans was not considered expertise:

> I let them go around and be experts. I told [the conservation director] that there's no point in keeping Ching and Ti in here. They say they can't control her. I say it's better for them to have their babies out in the wild [meaning free-range or bebas]. It's dangerous to have babies here. Are they going to take responsibility for the babies dying? Do they have the medicines in the vet clinic for them? [Our silence served as a "no."] Lin [the junior corporation officer on site] wants to keep

them here and have the babies here. She says we have to take care of the babies, to separate them [sighs in exasperation]. But if born in the wild, there's no need to separate them. If here, they have to be separated because they have nowhere to move. They don't want to take responsibility so they say that the other doesn't want it. But they're not going to bite anyone. You saw me in the enclosure with James. Let them be experts, with their education, and make the decisions—even if the babies die. I just keep quiet. (field notes, January 15)

Expressing his irritation about the orangutans' situation under the "rule of experts," he accused the corporation managers of not being responsible (Mitchell 2002). Layang's comment conveyed two senses of responsibility. The corporation managers seemed to lack the ability to offer material support in the form of medicine. Second, they refused to respond to what the orangutans themselves conveyed through their bodies. Layang often touched James through the bars and would place a hand on his belly or his hand. Otherwise, James tended to avoid physical contact and kept eye contact with humans for only brief moments. If there were too many humans within the night house, James would refuse to go back inside. By reminding me of how I saw Layang in James's enclosure, he was letting me know that such signs should be patently obvious, yet they either weren't to corporation managers, or they were simply ignored by them.

The death of the crocodile Stampin also serves as an example of how expertise is enacted through the performance of makeshift laboratory science. With the skin on its back opened to reveal pale flesh, after a scalpel parted that flesh to reveal part of the crocodile's spine, and with a chisel having been driven into it, Layang, Tom, and the others carried it over to the courtyard of the quarantine. It happened to be outside the window where my adoptive mother 'Kak cared for the infant orangutans. Tom's earlier concern over what he had called "ethics" matched ethicists' concerns in the analytic, British tradition: it was linked to an animal's ability to feel pain (Bentham et al. 1996; Singer et al. 1975). Likewise, Tom performed a European tradition of scientific knowledge production rooted in comparative physiology (Cole 1944; Park 2010). Like a sixteenth-century man of letters, he cut open the deceased body as a means of study. Tom dissected the crocodile's corpse over the next week. Later, 'Kak told me that she didn't understand why Tom would want to do that. She told me that it stank and it made her want to vomit. As we will see in the next section, the offense caused by the dissection of the crocodile's body may not have been about the smell of rotting flesh alone.

Crocodile Ancestors, Sun Bear Omens, and Orangutan Spirits

Wrestling crocodiles, facing down a sun bear, and coming close to an orangutan: all are part of a day's work at the interface of ecological loss at Lundu Wildlife Center. Yet some of these activities were historically taboo. The work of Iban ethnographer, folklorist, and curator of the Sarawak Museum after official decolonization Benedict Sandin (1967, 1980) points to a rich history of animal omens and augury. His research in the 1950s primarily involved elder informants, some of whom were contemporaries to Victorian anthropologists. Nineteenth- and early twentieth-century British men contributing to the new discipline of anthropology, such as Edward Burnett Tylor, Charles Hose, and William McDougall, turned to Borneo to think about primitivity, animism, and totemism.[25] Victorian anthropologists saw Iban and other indigenous Sarawakian peoples as primitive, especially in their relations with and conceptions of animals. Flash forward almost a century later, and similar sentiments of hierarchy and inferiority infuse lowland, urban prejudices against upriver practices and beliefs that have constituted upriver life in Sarawak. This presents a question: To what extent do omens relevant to an upriver past carry on at the lowland city's outskirts in a conservation workplace?

The answer I offer cannot be definitive. Discussions of omens were noticeably absent in my everyday presence at Lundu Wildlife and at home at Kampung Mohon. 'Kak, who at the age of thirty-four was only four years older than I was, once said, "The old people are capable of saying that bird means this, that bird means that. They're all just birds to me!" 'Kak dismissed such beliefs as old and irrelevant, and yet it was clear that such sentiments were not widely shared among her kin and neighbors at the longhouse. When a teenage girl still in school conceived a child and refused to name the father, Apai Julai and others felt that sacrificing a pig was the only way to quell bad blood in the longhouse. So they did so upon the riverbank, early in the morning after a long and inconclusive discussion on the *ruai* or common veranda of the longhouse. When 'Kak's sister-in-law had a stubborn cough, she consulted the same elder who had carried out the sacrifice of the pig. She left the consultation with drawings upon her back. Such activities show that there are no clean breaks between the old and new; there are no simple divisions between Christianity and Iban traditional beliefs, just as there are no clean divisions between the bandar and the *ulu*, or city and upriver. It is for this reason that Iban omens and augury involving endemic animals could be useful for understanding a framework through which Iban workers today could perceive their work with the same animals. The extent to how these beliefs might frame encounters can only be speculated about, since understanding beliefs entirely relies on what others willfully disclose.[26] Thus, I offer a look into Iban linguistic, folkloric, and historical encounters with animals as a way to suggest what might belie what is now a shared interface of displacement.

The literature on Iban folklore and religion points to the ways in which animals and other aspects of the living world around them have been central to Iban culture. Encounters with forest creatures are important for interpreting one's luck or misfortune. For instance, Benedict Sandin (1980) writes of *burong nganjong laba*, an omen that signals good luck when an animal approaches from the front into someone's rice field. However, if the same kind of animal approaches from behind it is a *burong nyabok*, which portends bad luck and the possible life endangerment of a family member within a year. Another bad omen says that encountering a sambar deer or bear on the path to one's rice field foretells one's own imminent death. Omens have a specificity about place and context, often in relation to rice fields or walking to and from home, and yet they are open to interpretation.

The combined uncertainty and power of omens is conveyed in a report by an employee of the Sarawak Museum, Guan anak Sureng, who accompanied Barbara Harrisson during her field research trip in 1959–60 through the Iban country of Ulu Sebuyau, my adoptive family's homeland and Alfred Russel Wallace's stomping ground in the 1850s. There, Harrisson investigated wild orangutans in their forest habitats. Guan (1960) reports the story of Serai anak Sinjai, who about twenty years before was walking on her way to her kebun or farming plot. She saw an omen bird and prayed for a charm to help produce enough rice without too much toil. Returning home with a basket of cucumbers, she passed by the same spot where she saw the omen bird. At that spot, an orangutan suddenly embraced her from behind. She screamed and dropped the basket. The orangutan let her go, picked up the cucumbers, and climbed the tree. Longhouse residents from near and far came to her cries. They decided to kill the orangutan. The orangutan was invulnerable to their attacks, until it was rid of a charm that later could not be found.

The story of one killing of an orangutan at the hands of indigenous people seems insignificant when compared to the many orangutan deaths caused by European natural historians in the name of scientific collection. Alfred Russell Wallace (1986) shot almost every orangutan he ever encountered, save for an infant that died in his care. He alone is responsible for twenty deaths. Odoardo Beccari killed twenty-seven on his own and with the help of Iban guides. William Hornaday, who later became the first director of the Bronx Zoo, killed forty-three orangutans, also in the name of scientific knowledge production (Baer 2008).

Yet this story is striking because the killing of the orangutan that embraced Serai anak Sinjai is an anomaly in the literature on conservation in Borneo. Many point out that killing orangutans is an Iban taboo, that orangutans are sacred (Horowitz 1998; Wadley and Colfer 1997).[27] In the context of the spirit world among Ibans of Batang Ai, from which Apai Julai hails, every household interviewed by geographer Leah Horowitz in May and June 1996 claimed that orangutans were manifestations of their ancestors or *tua* (elders). As tua, ancestors look after their descendants and offer luck and divine intervention. It would be a violation to kill tua in any form. Horowitz's interviewees also identified crocodiles as tua, along with gibbons and barking deer.

Horowitz's research, seemingly unbeknownst to the author, resonates with ethnological research findings from the century before. In 1899, colonial officer Charles Hose, with W. H. R. Rivers's student William McDougall, conducted ethnological studies at the Baram River shortly after the region's "pacification," where "only three or four white men had previously been" (Hose and McDougall 1901; Stocking 1995: 109). Like the tua described by Horowitz, Hose and McDougall (1901) write that *nyarong* are spirit-protectors of specific individuals that can take physical form as animals. They explain that once a nyarong has been identified as having taken on the physical form of an animal, killing a member of that animal species became taboo for the protected individual and for the individual's descendants.

The veneration of nyarong made it a difficult topic to address as outsiders and anthropologists. Hose and McDougall write, "The 'Nyarong' is one of the very few topics in regard to which the Ibans display any reluctance to speak freely. So great is their reserve in this connection that one of us lived for fourteen years on friendly terms with Ibans of various districts without ascertaining the meaning of the word *Nyarong* or suspecting the great importance of the part played by it in the lives of many of these people" (Hose and McDougall 1901: 199). The wariness around disclosing information about nyarong was well founded. In the same publication, Hose and McDougall describe how they commanded their Iban informant and research assistant to shoot the nyarong of his grandfather. The authors report: "Angus himself once shot a gibbon when told to do so by one of us. He first said to it, 'I don't want to kill you, but the Tuan who is giving me wages expects me to, and the blame is his. But if you are really the 'Nyarong' of my grandfather, make the shot miss you'" (Hose and McDougall 1901: 201). He shot at the gibbon three times and missed each time. On the fourth try, he shot and killed another gibbon.

When told to kill the gibbon, "Angus" in 1899 was forced to reconcile the demands of wage labor with the demands of adat and tradition.[28] More than a hundred years later, when workers perform their labor, it is difficult to ascertain whether or not omens have significance at Lundu Wildlife. Sandin's *Adat and Augury* (1980) notes that a common solution to encountering a bad omen is to stop work, from one day to as much as a week. To make good omens have power often requires the same solution. Such a refusal to work in order to pay respects to omens is antithetical to wage labor, when one no longer has autonomy over one's own time and often has little choice over one's activities.

Horowitz as well as Hose and McDougall point to the shift in which Christianity and wage labor threaten traditional ways. Horowitz writes: "Because of the influence of Christianity, many people are no longer afraid of punishment from the spirit world and so disregard ritual prohibitions" (Horowitz 1998: 383). Nearly a century before, Angus was forced to disregard ritual prohibitions. In the name of science, Hose and McDougall, the latter of whom would go on to chair Duke University's Psychology Department, tested the point at which "Angus" would draw a line. Hose and McDougall's expertise here relied on literally killing relations between their research assistant and his grandfather's nyarong and commanding new relations of wage labor for scientific knowledge production.[29]

Conclusion

With this historical context in mind, we can imagine what it might have been like for Ren, who had to tie up the crocodile and participate in its prolonged death, just as we might be able to better understand 'Kak's disgust at having to be confronted by the crocodile's corpse. Knowing that it is better in more ways than one to face an animal than to be sneaked upon, we can perhaps better understand why Layang faced the bear head on. Wage labor is central to the shared interface at Lundu Wildlife; the trace of the spirit world is perhaps another.

Gaining an interface of loss was made possible through the shared experience of displacement. Like Bako, animals that have arrived at Lundu Wildlife have lost their habitats and have nowhere else to go. Ngalih and Ren are two of many who have lost their former subsistence livelihoods in the highland interior of the state and have moved to the outskirts of the city, where conservation work at Lundu Wildlife is one of many possible ways of earning a living or cari makan. As we have seen, multiple elements shape the face-to-face encounters between displaced wildlife and their caretakers: displacement, wage labor, and shared connections across bodily forms. What possible futures are tenable under these circumstances? This last question guides the final section.

PART III **FUTURES**

Chapter 5

ARRESTED AUTONOMY

Climbing through the canopy from the north of Batu Wildlife Center, four-year-old male juvenile Umot or five-year-old female juvenile Nini could appear as early as one o' clock, a whole two hours in advance of official feeding time. Sometimes Nadim or one of his colleagues would end their own lunch break early to feed them. Sometimes, however, these young juveniles would come to the feeding platform after most of their conspecifics had moved on. Umot's mother Norma recently gave birth to her fourth infant and rejected him during her pregnancy, just as she had done with all her other offspring on site. This was an earlier age than observed among wild populations: infancy in general lasts for about three years, followed by a juvenile period in which they continue to accompany their mothers and continue learning (Wich 2009). Around the age of seven or eight, they begin to live apart.[1]

The recently weaned juveniles Umot and Nini are officially considered "independent" (bebas). Independence from their mothers means that they were even more dependent than other orangutans on the feedings administered by wildlife center workers. "Independence" entails competing with bigger and faster-moving orangutans for food handed to them by human workers. Such independence is in fact dependence. Though they had attained a freedom of physical mobility, that freedom is constrained by their pressing dependency on humans providing them the means to survive. Because of this need, they could not stray too far from the courtyard where they were fed—and if they did, they had to eventually return. Food, space, and acclimation via the work of care keep the orangutans' autonomy in a state of arrest. As we will see in this chapter, the work done to foster their independence instead fosters their dependency. This inertia is arrested autonomy.[2] Everybody and every body in Sarawak's wildlife centers, especially the orangutans and their Sarawakian caretakers, experience the unrealized promise of decolonization and the unrelenting grip of a colonial present.

I offer autonomy here as a translation for the word *bebas*, a physical kind of independence for which workers aspired for their charges.[3] Autonomy understood through the Malay term *bebas* is at once personal and political. It is especially political in respect to Sarawak's relationship to the nation-state of Malaysia. As a condition of Sarawak's incorporation into Malaysia, it is officially "semi-autonomous." At first glance, it would seem that the semi-autonomous nature of the state carries over into the realm of possibilities for wildlife, who are able to be physically autonomous within the constraints of designated wildlife centers. However, these conditions of captivity are more insidious in that the effort to make them eventually independent can only guarantee their perpetual dependence.

I see that the autonomy at stake in these wildlife centers offers another way of thinking about independence that differs from its intellectual history in liberalism and later in cybernetics. In this I am not alone: I join feminist philosophers Catriona Mackenzie and Natalie Stoljar (2000) in their appropriation of autonomy as "relational autonomy." I also join geographers Rosemary Collard, Jessica Dempsey, and Juanita Sundberg (2015), who see animal autonomy as a strategy for "abundant futures." Thinking of animal self-determination, or what Lori Gruen (2011) understands as the autonomy of being free of others' impositions, depends on acknowledging orangutans as selves among many other kinds of selves whom they encounter (Kohn 2007).

The state of arrest at Sarawak's wildlife centers is not just in the arrested development of subadult male orangutans on site, whose co-presence with the sole adult male suppresses their hormones and the secondary sexual characteristic of fleshy cheek pads (Maggioncalda et al. 2002). Nor is it just in the criticism of arrested development expressed by Iban scholars, who point out that Ibans and other Dayaks in Sarawak have been purposefully held in a state of arrested development ever since the nineteenth century and the formation of Sarawak as a state (Dimbab et al. 2000; Jawan 1991, 1993, 1994). Nor is arrested autonomy merely the fact of being constrained and held captive within the walls and boundaries of the wildlife centers. It is all of these elements and more.[4] The concept of arrested autonomy helps me understand how the end goal of self-determination comes to justify the means of indefinite constraint, even as those constraints further diminish the possibility of ever realizing that eventual independence. The paradoxical conditions of indefinitely deferred independence apply to both Sarawak's people and orangutans: thwarted freedom is a condition shared between almost everybody and every body in Sarawak.

Sarawak's people and animals survive in an extractive economy with unequal exchange, in which a few people have the freedom to move both themselves and the world around them as they please. In this chapter, I explain how arrested autonomy is a shared state between orangutans and the Sarawakian people who care for them. I locate arrested autonomy in the interface between orangutans, the people who care for them, and the political state that defines the space of their relations. My purpose here is to show how different kinds of Sarawakians at home in Sarawak, who through their relations with one another make Sarawak the place that it has become, share the politics of decolonization.[5]

The challenges that apparently free but actually captive animals face do not serve as mere allegories of Sarawak's century-long history of autonomy and despotism as a personal kingdom of a British subject, then a British crown colony, and finally a state in the federation of Malaysia. This is not just a parallelism. It is a story about real animals facing the existential threat of extinction, who, along with their human caretakers, are all too familiar with the violence of displacement for capitalist consumption of land and natural resources. This story folds into Sarawak's history of forced dependency as it is cast through the language of and hope for eventual independence that is experienced by most Sarawakians as perpetually deferred.

Understanding the problem of indefinitely deferred independence as it is shared among all kinds of bodies in Sarawak requires understanding the terms of freedom and independence available in the colloquial: the adjectives *bebas* and *merdeka*, and their respective noun forms *kebebasan* and *kemerdekaan*. The concept of bebas offers a feeling of physical freedom that is lacking in the English words *autonomy* or *independence*. A similar sense of self-directed material, physical freedom is expressed in Iban, which is in the same linguistic family as Malay: *'ati diri*, or the sense of being guided by one's own heart (or, anatomically speaking, one's own liver). Thinking through the concept of what it is to become bebas compels us to think about autonomy in another way, different from its Western civilizational contexts of analytic philosophy and cybernetics. Thinking of autonomy through translation, even knowing that translations obfuscate as much as they clarify, shows us that autonomy in all of its iterations remains relational and interdependent, not isolated and autological.[6]

This chapter illustrates the political possibilities that appeared in three different moments I experienced in which Sarawakian people empathetically spoke of how the animals they cared for suffered through the same structures of exploitation. This was apparent at Batu Wildlife Center, when Nadim shared the pity that he feels for the female orangutans under his watch, and when he and his colleagues stood and watched a large and healthy tree that survived the abuse of orangutan locomotion be felled for the sake of an additional toilet for visitors. It was apparent at Lundu Wildlife Center, when Layang shared his pessimism about conservation. It was also apparent at Lundu Wildlife Center, when one merely looked at the walls of the enclosures. The walls that literally forced orangutans to stay within their place were made from what amounted to hours of coerced labor. Structures of confinement and arrest were at times all too literal, and almost everybody (and every body) could feel it. I conclude this chapter by thinking through the conditional possibility of autonomy and what that would look like if it were ever to be realized.

"Free but Living in Fear"

As a prepubescent female orangutan, five-year-old Nini's independence was marked by the risk of forced copulation. Orangutans, as I explained earlier, are sexually dimorphic, and so male bodies are bigger and stronger than female bodies. This sexual dimorphism allows for forced copulation to be a typical behavior for the species. Orangutans that are Nini's age usually still accompany their mothers up until the age of about eight years. Without the protection of her mother, whom she would be able to cling to and escape with quickly, she was subjected to copulation by males who have the limb span and strength to outmuscle her and restrict her movements. Her prepubescent age did not seem to matter to sexually mature subadult males nor to some humans witnessing such moments when they occurred.[7]

Both Zeb and Nadim often used the English word *pity* to describe this situation. For Zeb, it was a pity to see newly weaned infant orangutans arriving at the feeding platform on their own, independently, only to find that no food was left. For Nadim, feeling pity was his response to the expressions he perceived on the faces of female orangutans at the wildlife center who were trapped in circumstances that were radically different from wild conditions:

He said, "six square kilometers is not enough. In the wild, there's lots of trees, lots of space. Here, it's six kilometers and not enough. Here, they meet every day! In the wild, they [males and females] meet in a year or once every six or seven years . . . they [female orangutans] may be free, but living in fear."[8]

I asked, "Bebas?" testing a word I had repeatedly heard at Batu Wildlife Center that had not come up in my Indonesian and Malay language courses.

Nadim answered, "*Bebas, tapi takut*," literally, "free but fearful." He continued, "I pity them when I see their faces. It's only the males, when you see them, they're happy."[9]

The Malay concept of bebas is significant. The Oxford Fajar 2006 edition translates the term as follows: "free; not a slave, not in the power of another; having freedom; not fixed, able to move; without, not subject to; not occupied, not in use; lavish; clear; free from doubt, difficulties, obstacles, etc.; carefree; light-hearted through being free from anxieties; footloose, independent; without responsibilities; not dependent on or controlled by another person or thing." This definition carries many meanings, coincides with other terms in Malay, and requires a deeper investigation into its usage in the past and present.

For nineteenth-century figures of the region, such as Munshi Abdullah, a translator and author from Malacca on the Malayan peninsula who was posthumously named the father of Malay modernism by historian Anthony Milner; Sir Stamford Raffles, founder of Singapore and the sole inspiration for the first White Rajah James Brooke's desire to take personal control of Sarawak; or William Marsden, author of one of the first Malay–English dictionaries, published in 1812, the ideals of freedom and liberty were conveyed through the Malay idea of merdeka (Milner 1994; Reid 1998).[10] The historian of the British Empire David Armitage (2000) has argued that freedom and liberty were key ideologies that distinguished British imperialist rhetoric from other competing empires. While the Spanish and Portuguese were depicted as relying on brutality, and while the Dutch were dismissed as being interested only in economic exploitation, British imperial rhetoric after the abolition of slavery rested on the idea of spreading liberty and uplift through the eradication of piracy and the facilitation of free trade (Armitage 2000; Reid 1998; Tagliacozzo 2005).[11] Like many abstract terms in Malay, merdeka has roots in Sanskrit and in the ancient Srivijaya and Majapahit imperial courtly and elite cultures that connected much of Southeast Asia.

Merdeka, with its ancient imperial contexts consistent with British imperial ideas of freedom, differed from the commoner roots of the word *bebas* (written by Marsden as *bibas*). The use of Malay as the vernacular trade language throughout the archipelago spread the latter term; Orientalists like John Crawfurd (1856) thought the term originated from Johor on the peninsula. Bebas seemed to represent something more excessive, even "wild," distant from "civilized" courtly culture, further from the ideas of liberty conveyed in the writings of John Locke and John Stuart Mill, and further still from the stately sense of the term *merdeka* (Kirksey 2012; Rutherford 2012; Steedly 2013).[12] Marsden's translations of the terms *free* and *liberty* in the 1812 edition of *A Dictionary and Grammar of the Malayan Language* convey this: "Free (manumitted) *mardika*; (on equality) *sama rata*; (unrestrained) *bibas*. . . . Liberty (enfranchisement) *ka-mardika-an*; (permission) *mohon, bibas* (1984: 451, 482).

While merdeka and its noun form kemerdekaan are connected to emancipation and enfranchisement, key ideals in British liberalism, bebas is associated with license and lack of restraint, which does not sit easily with British imperial sensibilities. Crawfurd, resident of Singapore and an early scholar of Malay and Javanese, defined the term *bebas* in his *Grammar and Dictionary of the Malay Language* from 1852 (p. 28) as "free, unrestrained; license, liberty, permission." Shellabear's *An English-Malay Dictionary* from 1916, and used by expatriates in Sarawak during the reign of the second White Rajah Charles Brooke, translated bebas (p. 212) as "free" and "not under restraint."[13] It is a feeling of freedom more physical or literal.

More contemporaneously, the term *bebas* arose when Aihwa Ong studied young Malay women working in factories in the 1980s, as they expressed what she describes as "autonomous female agency" (Ong 1990: 268). Challenging traditional hierarchies of gender and sexuality within Malay villages in rural peninsular Malaysia, Ong (1990) described them as "unrestrained (*bebas*) by family guidance in relations with men." The adjective *bebas* is not as pejorative as *liar*, which translates as wild, and references both wildlife, *hidupan liar*, and wild behavior, *berlakuan liar*, often exhibited by people at wild and sexually uninhibited parties, or *pesta liar*.

Today, the term *bebas* comes up when discussing free trade, as in *pasaran bebas*. For instance, the head of the Dewan Perdagangan Islam Malaysia (DPIM) in 2014 publicly stated, "*pasaran bebas* [free market] *hanya menguntungkan golongan kaya* [only benefits the rich]" (Zain 2014). The term *bebas* also appears in the legal term for acquittal, *pembebasan*. When the former chief minister of Sarawak, Taib Mahmud, resigned to become governor of Sarawak and thus its official head of state, opposition party representatives in Sarawak suggested that he should be free of corruption charges, *bebas rasuah*, before assuming the position, which is equivalent to the position held by sultans on the peninsula or by Queen Elizabeth of the United Kingdom (Davidson 2014). Additionally, Wikipedia in Bahasa Malaysia is known as the "ensiklopedia bebas," which is not free in the sense of being free of cost, which would be the loan word of *gratis* from Latin via the European trade languages of Dutch and Portuguese. Rather, this is a freedom of unrestrained circulation of information, in contrast to the censored state-owned media, although that freedom is questionable when considering the limited number of contributors creating Wikipedia's content.[14]

Nadim's observation that the female orangutans "may be free [bebas], but live in fear" reveals much more than a mere report on their conditions of mobility. His explanation expresses empathy through identification. It was upon sight of female orangutans' faces that he felt enough empathy to imagine female orangutans' constant fear. By explaining that his feelings arose when gazing at their faces, he located the evidence for the need to feel pity. He is not necessarily more sentimental than another onlooker; rather, his explanation marks him as an attentive observer. Nadim is both an empiricist making observations and open to affective contact produced through his co-presence with orangutans and in the space between his place on the ground and the apes' places up in the trees.[15] The distance between ground and canopy guarantees that he can do nothing but either watch or ignore prepubescent forced copulation.

Historically, the term *bebas* was "pedestrian": it was a term "always available to indicate freedom of movement" (Reid 1998: 51). This pedestrian sense underlying bebas is apparent in Lockard and Saunders's (1972) *Old Sarawak: A Pictorial Study*, a book sponsored by the state governments of Sarawak and Sabah less than ten years into their incorporation into the federation of Malaysia and published in the wake of the first New Economic Policy in 1970, which was meant to sponsor national unity in multicultural, multilingual Malaysia following the May 13 race riots between Malays and Chinese on the peninsula. Published in four of Sarawak's dominant languages at the time—English, Malay, Iban, and Chinese—the book describes the Brooke dynasty's rule over Sarawak as "indisputably autocratic. And yet it was popularly based" (Lockard and Saunders 1972: 129). This was conveyed in how "the Rajah and his European officers moved amongst the people they governed with freedom and ease," in which freedom was translated as bebas (Lockard and Saunders 1972: 130).

The physical freedom James Brooke, followed by Charles Brooke, and then later Charles Vyner Brooke enjoyed in walking around Sarawak was, of course, afforded by the absolute political power they each in turn held over the land as the ruling rajah.[16] While each of the Brooke rajahs could be bebas, their subjects could not.[17]

Nadim's statement that Batu Wildlife Center's female orangutans are "bebas, tapi takut," or free but fearful, while the males were the only ones happy conveys the idea that male orangutans could roam with freedom and ease in the way that the White Rajahs could roam with freedom and ease in their Raj. However, it was not true. Eventually, subadult males like Aqil would grow older, just as James did. James challenged the massive adult male Lucas in a series of ferocious fights that led to James becoming blind in one eye. The fighting between them appeared so disturbing to visitors that James was sent away to Lundu Wildlife Center, where he is locked up, spending his days in an enclosure and his nights in a cell.

The emphasis on freedom of movement in the term *bebas* suggests that the idea of kebebasan or being bebas has material, physical forms that get lost in the English word for autonomy. There is a political depth in the terms of freedom, of being bebas, that is not offered by the concept of wildness or liar. This context is certainly not free of cost or gratis. This freedom is not the lofty freedom of the nation-state following official decolonization, or of merdeka. The autonomy striven for is the physical ability to move about in the world without fear.

"When There's Money, There's Development"

Batu Wildlife Center occupies land that had been acquired by the first White Rajah, James Brooke, late in his administration. The 1863 Land Order decreed that all "unoccupied and waste lands" were the property of the government, which meant that all land determined to be unoccupied by natives belonged to the Raj and thus was to be administered by the rajah (Kaur 1998).[18] This served to dispossess nomadic people like the Penan and has curtailed the activities of swidden agriculturalists like the Iban in the interiors of Sarawak and the Bidayuh, who have historically lived in the area around Batu Wildlife Center.[19] In the 1920s, Batu Wildlife Center was officially founded as Sarawak's first forest reserve for forestry research, during Charles Vyner Brooke's administration as the third White Rajah. His administration began in the midst of a boom economy resulting from worldwide and local demand for timber and nontimber forest products, when the value of timber exports rose exponentially, from S$153 in 1870 to S$30,144 in 1920 (Kaur 1998). He created the Forest Department to answer these demands and appointed its leader, J. P. Mead, as Sarawak's first Conservator of the Forest. Mead's Forest Rules of 1919 established the boundaries for Batu Wildlife Center and also established the system of permits and licensing for the collection of all forest products, including nontimber products traditionally harvested by Dayak smallholders, including Ibans and Bidayuhs (Smythies 1963). The decree also established a Communal Forest, in which forests for natives were limited to subsistence consumption alone. Participation in the economy through cash crops was heretofore regulated by the state through permits, thus putting smallholders at a disadvantage and larger-scale capitalists at an advantage.[20] The Forest Rules of 1919 bound the limits of indigenous people's mobility and territory within Sarawak. Beginning in the 1970s, these rules directly impinged on the mobility of the orangutans that found themselves at Batu Wildlife Center.

Since the 1970s at Batu Wildlife Center, both local urban and foreign visitors could experience wildlife that they would otherwise never encounter in the wild. When it first opened, one could only get to the wildlife center if you walked three kilometers up the dirt and mud path that weaved up, down, and around the hilly terrain. By the 1980s, one had to drive up a single-lane paved road that eroded a bit more with each rainy season. By the time I did my research between 2008 and 2010, they had stopped repairing the road. Visitors in vans driven by tour guides ride up and down the curvy three kilometers. At one stretch, the asphalt of the shoulder and part of the lane had crumbled down the eroded hillside. The managers of the park applied for a grant under a federal program that allotted money for the improvement of national parks, and Batu received money for infrastructure improvement for the road. But the funding wasn't enough to repair it, so they looked at the budget and decided what to do with the funding: they decided to build new toilets.

To do this, they cleared part of the forest behind the park headquarters. Peter, the park manager as the PIC or "person in charge," had not consulted Nadim the park ranger, who was under him in the company hierarchy. As a park ranger, Nadim's job entailed feeding the orangutans and managing the physical distance between orangutans and human visitors inching ever closer to take snapshots and gain physical proximity to members of a famously elusive species. Nadim would have told him that the area selected for the new toilets was used as a corridor by the orangutans in the wildlife center—that the area connected them from the northern section of the park to the main feeding platform. He would have told the PIC that the older toilets should just be renovated and that they should build a fenced and covered perimeter for the parking lot. However, Nadim was in no position to give his superior his unsolicited advice.

It took a week to clear the forested patch around the headquarters. As the very last visitors from the morning feeding would leave, the contracted laborers would rev up the excavator's engine and haul away the timber cut by hand to the pile of debris. One day when the project was well under way, Nadim and I were sitting at the threshold of the building. With its corrugated tin roof and area open to the veranda, the building was hot without the shade of a tree canopy. The excavator and its operator did the work of ripping down tree trunks and flattening the hill beside the park office. Over the noise of cracking open a large living tree with an excavator, Nadim shared his thoughts with me:

It's like what I told you yesterday: when there's money, there's development. It's now already hotter here without the trees. Who knows, in 20 years time, they might want to build on that side and say here's no good. [He gestured toward the other side of the single-lane paved road, where the trees of the forest provided pathways, cover, and areas for the more arboreal orangutans on site.] The orangutans have less trees. Already there's not enough space. We are forcing them to stay. They are forcing them to stay—because of food and because of space. If I was with my baby, I don't know what I would do. I think that's why those that are lost have to come back. Wani was gone for 2 weeks and now she's back. I think she tried, but not enough food and nowhere to go. She knows food is here.[21]

Nadim goes further than just separating himself from the position of managing a diminished and displaced population of orangutans. On a discursive level, he identifies with an individual member of the population he is managing: "If I was with my baby." He claims a first-person identification as the one directly experiencing loss. His identification also crosses a gendered identification because he is identifying himself with Wani's position as an orangutan mother. Doing so shows his willingness to imagine the subjective experience of the orangutans he "manages" as the park ranger.

Nadim criticizes the destruction of a forested area for toilets as an example of exploitative development. He does so openly with me, an outsider, yet he contains his criticism when talking with his supervisor, the PIC.[22] I was not in a position to ask pointed questions without risking my ability to continue my field research. Yet I was compelled to ask the PIC why such a construction job was necessary. He explained that the new toilets had to be built because the location of the older toilets up by the parking lot was too "inconvenient" for visitors. The PIC left it at that.

Nadim is forced to keep up a relation of forced docility and to collude with an orientation to ecology that sees value only in human value, whether that is the human aesthetic value of free-range freedom or the human convenience of an additional toilet. The "human" here in human toilets does not stem from some kind of universal human value. It comes from a particular modernist ideology that aesthetically values forests in the midst of their destruction, orangutans on the brink of their extinction, and the appearance of their freedom in the midst of their active enclosure. This exemplifies what Tim Choy calls "anticipatory nostalgia," akin to Renato Rosaldo's "imperialist nostalgia" (Choy 2011; Rosaldo 1989). It does all of this while also prioritizing the conversion of an already very small space for orangutans for a mere convenience.

When considering the further diminished space experienced by these orangutans, Nadim says, "we are forcing them to stay." As a ranger, he represents the Forest Corporation and he uses the first-person plural pronoun. But then he refuses to implicate himself; he quickly corrects himself: "They are forcing them to stay." He then further separates himself from the position of manager and slips into a discursive identification with managed wildlife. By saying that he would not know what to do, he seems to suggest two possibilities. Either he rejects the ability to slip into an orangutan's subjectivity, or he acknowledges the lack of viable action to be taken. The answer seems to be the latter, since he elaborates on Wani's position within the wildlife center: she tries to be independent and self-sufficient within the allotted space but the physical, material constraints force her to return to seek out food at the center, where the workers distribute it on the understanding that large groups of people will watch and take photos.

Nadim recognizes the coercion inherent in the caring act of offering food support to free-ranging orangutans that have "nowhere to go." He himself is stuck, unable to criticize the development around him and unable to provide anything more than food. By providing "bare life" for orangutans like Wani, he attests to the degree of arrested autonomy that he experiences—both by working with her closely and seeing her struggle as well by his inability to provide more than mere subsistence (Agamben 1998). Nadim is forced into a position of complacency, having to watch two adolescent male orangutans forcibly copulate with prepubescent Nini and having to hand food to Wani whose attempts to become autonomous are futile.

What can Nadim do beyond feeding Wani and all the other orangutans? He shares his opinions with those who listen. His words seem to be steeped in a tradition of Malay political critique, one that he roots in Islamic political vocabulary. When talking about the pending extinction of orangutans, Nadim said, "It'll be fine in your lifetime, in my lifetime, after we're dead. It's for this young generation, these young kids and their children. It's up to God because we won't be here. But God gave us a brain to use and to think."[23] Nadim's language resonates with Malay political discourse reminiscent of the journal *Al Imam* from the early twentieth century in

Malaya that critiqued the despotism and hegemony of the *kerajaan* in its original sense, before the term came to reference government at large, back when it referenced only royal elites. This political tradition deployed a religious imperative in which one's reasoning (*akal*) leads to ethical action (Milner 1994).

Nadim's ethical dilemmas and critiques of the situations he is forced to witness reveal how the wish to act autonomously extends beyond acting for oneself alone. He wishes to act on behalf of others and to advocate for their needs, namely for the orangutans under his care. Yet the job that allows him to connect with his charges also prevents him from exercising what that care should be in his eyes. He recognizes that Wani's independence is simply unattainable, though she actively seeks it. His job, then, is to fulfill an impossible task. This impossibility demonstrates how Nadim experiences arrested autonomy as an employee in his particular workplace and the relations that are allowed to happen: care for mere survival alone.

"How You Do Conservation"

"Animal management" is dangerous work, especially when it involves a species not evolutionarily inclined nor bred to be in contact with humans. Workers are vulnerable to carnal violence from orangutans, who have massive incisors adept at chewing durians and bark. Nadim himself was bitten once. As he described it, Deh ripped into his leg like it was sugar cane. The possibility of violence is real since these particular individuals are inclined to stay near humans instead of keeping their distance. Those who experience this possibility six days a week repeatedly say that it is never the animal's fault when an animal bites. The responsibility falls on the human being bitten.

Yet the term *animal management* is a misnomer when applied to the work of Nadim, Layang, and 'Kak. Everyone at the Forest Corporation who ever worked with animals during the time of the Forest Department repeatedly said the same thing to me: "You can't control the animals; you can control the visitors." Visitors would repeatedly breach the recommended distance of five meters, wrongly assuming that either the animals were tame and docile or that the workers could protect them. Workers and managers alike emphasized again and again the impossibility of their official task of "controlling" or "managing" animals.

Despite the potential of losing flesh, fingers, or bodily integrity as a result of a bite, the keepers stay committed to their work. Zeb proudly speaks of himself as having served the state as a civil servant in the Forest Department. He and many of his colleagues worked during the time of the Forest Department and grew up in the early years of the nation. During interviews with me, they would sometimes say that they should write a book about their experiences. Their wish resonated with the biographical tradition of "great men" who have helped to establish Sarawak: the White Rajahs who ruled autocratically for a century; Tom Harrisson, the famed war hero who became the curator of the Sarawak Museum; Chief Minister Abdul Taib Mahmud, whose term that began in 1981 made him the longest-serving chief minister in all of Malaysia and one of the longest-serving politicians in the world; civil servants working for the district offices during the days of the British crown colony; and Tan Sri Ong Kee Hui and Tan Sri Stephen Yong who cofounded the Sarawak United People's Party, Sarawak's first political party, which shifted its reputation from harboring communist sympathies to being the favored partner in Sarawak's dominant political coalition since independence.[24] In the same breath of pride over having had a government job for most of his working life, Nadim's colleague Zeb would share his dream of becoming rich through MLM—multi-level marketing—of getting rich through worldwide business contacts and of being able to afford everything that he needs or wants. Zeb, from his current position framed by the vestiges of the civil service, earnestly harbors the fantasy of what anthropologist Aihwa Ong (1998) calls "flexible citizenship," of being a successful neoliberal agent, getting money through speculation rather than actual goods, products, or substances. Cynically, Nadim tells me that "management is just interested in their pension." Nadim recognizes a hierarchy of labor between management and employers. All of them point to a frustration with the current circumstances of their jobs.

Most cynically, Layang at Lundu Wildlife Center, in one of his more depressed and frustrated moments, complained about the state's moral panic about indigenous poachers of orangutans, when the real problems were instead massive monocultural oil palm cultivation and dam construction. He said:

> If I had the money, Juno, I'd open an oil palm plantation. Just spend money for 2 years. And then after that, I can sit in the office. But I'm

not rich. You just wait. If you have the time, you should go to Mulu Dam. It's all oil palm. They started oil palm there 2 years ago—and it's still opening for oil palm. Whose habitat? Langur, pigtail [macaque], proboscis [monkey, all of which are totally protected species in Sarawak]. Where are they now? Right now my idea is to be friends with royal people. It's just all about money. That's how you do conservation.[25]

Layang passionately criticizes the exploitation of elites through agricultural business. His quip about just sitting in an office was a jab at the "Ali Baba" way of doing business following the May 13 riots in 1969. Malaysia responded to the peninsular race riots by instilling a "Bumiputera Policy" in the New Economic Policy as a way of encouraging Malays and other "sons of the soil" to go into business. The rules meant that joint ventures had to have 51 percent Bumiputera ownership. An "Ali Baba" business is when a non--Bumiputera partner runs the operation or when a Bumiputera-owned company gains a contract that would then be subcontracted to a non-Bumiputera company. Without capital, Layang could not participate in such a program of economic uplift. Layang expressed the commonly held opinion that Ibans and other Dayaks in Sarawak and Sabah were second-class citizens, entitled to certain rights on paper but denied those rights in practice.[26]

Layang's comments were made not long after a crown prince, a son of one of the nine royal rulers on the peninsula, heard about an orphaned gibbon that was brought to the wildlife center. He sent a representative to personally check on the gibbon's status. The Forestry Corporation formally received the visitor, sending an entourage in a fleet of company cars.

The formality of royals was alien to Layang since the significance of the institution is concentrated in the peninsula. Sarawak doesn't have a sultan who acts as head of the state; its contact with the Brunei sultanate was peripheral at its coasts more than a hundred and fifty years ago. Sarawak's head of state, a position currently occupied by the former chief minister, is a civilian governor. Although the rajahs adopted the signs of sultanate power early in the administration of Sarawak, they maintained relations of personal rule: Charles Brooke heard complaints from his subjects every afternoon at the Astana, and officers were instructed that their subjects "are *not* inferior but they are different" (Kedit 1980: 59). In ethnological texts, Ibans are famous for being egalitarian.[27] Layang regarded Malay royalty without reverence; in his eyes, they were indexes of wealth that he could never have.

The distance from which Layang regarded the peninsular royal institution reflects the geographical distance between Sarawak and Malaysia. Before jet travel, transportation beyond Sarawak was mostly directed to and from Singapore, the southernmost hub of the Malayan peninsula and the port of Malaya closest to Sarawak. A boat journey between Kuching and Singapore took two weeks in the nineteenth century (Payne 1986). In 1960, the same journey took Barbara Harrisson and two caged orangutans about four days on a P&O liner (Harrisson 1987).

The distance of Layang's attitude toward royalty also reflected historical differences between Sarawak and Malaysia and Sarawak's contingent incorporation into Malaysia. Despite Charles Vyner Brooke's promise of eventual freedom in September 1941, Sarawak was instead transferred to British control after the Japanese occupation. In a public address on February 6, 1946, Brooke stated, "Your future happiness lies within another realm. There shall be no Rajah of Sarawak after me. . . . This is for your own good: By Royal Command" ("Rajah Agrees" 1946). The transfer of power included a £1,000,000 trust fund for Brooke's family for their maintenance at prewar levels. It was taken out of Sarawak's £3,000,000 financial reserve (Kedit 1980; Payne 1986; Yong 1998).[28] His words were performative: there was no rajah of Sarawak after Charles Vyner Brooke.

In 1961, Sarawakians experienced another transfer of power at the highest levels. The prime minister of Malaya across the South China Sea expressed the wish to form a federation between Malaya, Singapore, Brunei, Sarawak, and Sabah. This declaration was made without taking any Sarawakian opinions into account. The Cobbold Commission was then appointed to assess the attitudes of Sarawakians about such a federation. They found that a third supported it without any conditions, another third was against it, and the remaining third was apprehensive and would support it if some conditions were met (Ongkili 1967).[29] The British and Malayan governments interpreted the findings such that a two-thirds majority supported the endeavor. The colonial government cracked down on opposition by issuing the Restricted Residence Ordinance in October 1961, which authorized the exile of critics of the government. This sparked a movement in Sarawak pushing for evaluation by the United Nations Decolonization Committee. It also led to a communist insurgency. The UN team visited Sarawak in April 1963 and encountered protests. Before they could even issue their report, the British Governor Sir Alexander Waddell named Stephen Kalong Ningkan chief minister of Sarawak, who then declared on September 16, 1963, that Sarawak had joined the Federation of Malaysia. Sarawak's incorporation into Malaysia has fostered relations of extraction. For instance, Sarawak retains only 5 percent of the revenues from its oil (Yong 1998). Sarawak, however, does retain autonomy in regard to immigration (He et al. 2007).

The history of Sarawak's transfers of power from the Brooke family to the British Crown and eventually to Malaysia shows that official "independence," or merdeka, in Sarawak is in fact its cession to another authority. Layang and his peers know that immense resources are at stake in such transfers of power, as they are in contemporary agricultural development and dam construction. As a wage laborer officially making 900 RM, which is supplemented by a subsidy from ENGAGE, the commercial volunteer company operating at Lundu Wildlife Center, Layang knows that his wages are insignificant in comparison. Knowing this, we can see that his complaint that it is "all about money" is a complaint about concerns regarding an individual animal's welfare without thinking about the conditions under which that animal came to need help. It is also a complaint about social station, in which a royal's concern about an animal far away is more important than concern of the handler who works with the animal every day.[30]

To Speak "White People"

Tom used the word *help* to describe his commercial volunteer company's role as an employer at Lundu Wildlife Center, understaffed since privatization. Help from Tom and the for-profit volunteer company he worked for was directed toward the animals and had implications for the people too. But what he construed as help meant maintaining arrested autonomy for both humans and orangutans at the site.

'Kak, who worked as the primary orangutan nanny, technically worked for Tom's company. 'Kak's story helps show how she experiences arrested autonomy, a state of dependence that is framed as a stepping-stone to what is in practice an indefinitely deferred state of independence. To describe her work, 'Kak would never use the term *nanny*. The closest Malay word to nanny would be *amah*, which means maid, and she would not use the

English term either. "Orangutan nanny" was a descriptor given to her, but not an occupation or any other noun that she professed. She instead used action words to describe her employment: "*jagah orang* [*utan*]" (caring or watching out for the person/orangutan).[31] Her work was a means of wage earning and not an identity-oriented calling or profession.

While conducting research at Lundu Wildlife, I lived with 'Kak and her family in the Iban longhouse just outside the front gate of the wildlife center. As I explained in chapter 4, Kampung Mohon longhouse dwellers had come from all parts of Sarawak in the early 1980s and pursued wage labor.[32] When the center was being built, 'Kak worked as a construction worker and helped build the orangutan enclosures. She worked along with the women of her longhouse in construction as subcontractors, as all of the adults of this longhouse usually did.

'Kak and her work party built two animal enclosures over the span of three months. After the job was finished, the *towkay* or company boss skipped town right after he underpaid them all: 300 RM for three months of work instead of 900 RM. 'Kak told me she couldn't go to the Labor Department to complain because their promised wages were too small to file a claim. Even so, she was resigned to the sense that nothing could be done. The very confines of Lundu Wildlife are indications of the confines experienced by local laborers. The site depends on the labor and compliance of 'Kak and her fellow villagers for construction projects, farming food for the herbivores on the site and selling it to the center, and carrying out basic animal husbandry for the animals in captive conditions.

'Kak's husband Ren has been an animal keeper since the center was completed in the 1990s. They selected him and many other men from the day laborers who constructed the park. After the Forestry Corporation took over the park and biodiversity conservation responsibilities from the Forestry Department, Ren no longer had access to overtime pay. So he and 'Kak had to become flexible when it came to cari makan, or getting food on the table for their family.[33] On days when he was able to get a better-paying construction job, 'Kak substituted for Ren at the wildlife center. At first, she didn't want to. She was afraid of the animals. She was afraid of the crocodiles, to which she'd have to toss dead chickens across a wire fence. She was afraid of the binturongs, black with yellow eyes, coarse fur, and a laugh that sounded to her like it belonged to somebody malevolent. But her husband said she had to do it because they needed the money. So she did.[34]

'Kak already had lots of animal husbandry experience before she became the orangutan nanny.

As the primary orangutan infant caretaker, she's in a slightly better position than before. She is paid 500 RM in cash per month, without documentation and without deductions for the Employee Provident Fund, which is the federally managed fund used for retirement, disability, and unemployment. Her wage is less than the poverty line of 800 RM, and that figure itself is questioned by economists as being too low (Mahavera 2010). At Lundu Wildlife, she's technically part of the staff hired by the British volunteer company. Tom hired her and other villagers from the nearest village as part of their effort to enlist "community support." As Tom had said to me once, "the key to conservation is employment." 'Kak joined the staff at the same time as Ngalih, her neighbor who was also her nephew-in-law. They started at the same wage of 500 RM. Ngalih's job entailed cleaning cages, feeding animals, working on construction projects, and helping, instructing, and entertaining the sets of monthly and biweekly volunteers who came from abroad.

'Kak's job was at first inside in the vet clinic, where the orangutan infants Sri and Lubok were kept, and later outside in the special play area for the infants. When she started, she took care of just one infant less than a year old, Sri, from 8 AM until 5 PM with a two-hour lunch break. Tom didn't believe in having orangutan infants wear diapers. Indeed, he hated the image of infant apes in diapers and found such images to be "anthropomorphically indulgent." He devised a system whereby the infant apes would nest in plastic hampers lined with a terry cloth towel. The infant would then urinate and defecate on the towel, and soiled towels would be replaced. Her job was to clean all the soiled towels, feed the infant, catch his feces and urine with a rag if it happened outside the hamper (which it usually did), and watch him. She had to make sure the infant wouldn't hurt himself by climbing and opening the cabinets, broken lab equipment, and refrigerator. When he climbed away too high from the basket, she would gently pry his limbs away and put him down at the bottom of the basket. After six months of this, another infant came and her work thereby doubled.

In January, Ngalih negotiated a raise. He was then making 1,000 RM to 'Kak's 500 RM. Yet 'Kak's work doubled and her wages stayed the same. She was caring for two infant orangutans at once for the same wages as caring for one. In March, we talked about the discrepancy in wages while she watched both of the infants in their special play area:[35]

'KAK: His salary's bigger than mine. He said 500 RM was not enough. He spoke to Tom. Tom added to his salary so it was 700 RM. He [Ngalih] said again it was not enough. From January, he received 1,000 RM.

JUNO: Did he ask for the 1,000 or did he just receive it?

'KAK: He asked. His salary was raised. It's not enough. I work a lot. My job is not just like this [she pointed to the orangutans behind her on the jungle gym]. I have to wash towels [on which the infant orphaned animals urinate and defecate], mop, sweep inside [the clinic cum nursery], and watch the orang.

JUNO: You don't want to ask Tom?

'KAK: No. I'll just wait for Tom. [silence]

JUNO: Ngalih got a raise because he asked. Why don't you ask?

'KAK [quickly]: Because Ngalih's capable of speaking "white people" [*orang putih*, which literally means "white people," is a referent for English language]. He's capable of talking about his salary, of needing it to increase.[36]

Speaking "white people" in this respect literally meant speaking English, but it also meant being able to negotiate one's own labor conditions. In her complaint to me, she never pointed out that her workload had in effect doubled. She instead talked about the range and number of tasks for which she was responsible and the length of time she had worked there at the same wage.

Wages were a very contentious issue. Often Tom would brag to volunteers and me that he took only 300 RM a month for his living costs. Yet he was not raising children, used the company van, had free rent, and preferred cheap Indonesian cigarettes and coffee over meals. Two others on site were employed by the British company and could speak English, but no one wanted to translate for 'Kak. Layang refused and said that the issue was too "sensitive," which in Sarawak is synonymous with political controversy and subversion. He explained that if he or anyone else were to ask Tom, the person asking would look like he's complaining that he doesn't have enough money either. When I asked Tom about her wages, he said: "'Kak has the easiest job in the world. I'd love to trade places with her!"

Tom's statement shows the lack of regard he has for her feminized activities of cleaning and caring for infants as work that deserves at least poverty-level wages. Perhaps to some outsiders, especially those coming from abroad, physical proximity with infant orangutans is worth poverty wages. 'Kak does not have the autonomy to be able to choose work according to whether it offers personal fulfillment. Without a raise, she has to depend on Tom for her livelihood. Wages for her are not a means of modern independence, but another form of dependence.[37]

Arrested Autonomy, Trans-Specifically

What happens when we see that attempts at gaining autonomy result in a state of indefinite arrest? In other words, when independence proves to be impossible, what alternatives are there? What would unrestrained autonomy look like here? This requires thinking through the idea of autonomy.

The concept of autonomy, rooted in the Ancient Greek *autós* (self) and *nómos* (law), traces its intellectual genealogy in liberal humanist philosophy and is crucial for thinking about continued struggles for self-determination in the postcolonial nation-state. Jean-Jacques Rousseau and later Immanuel Kant felt that autonomy was not unbounded freedom from any constraint, as it is too often characterized. For Rousseau, autonomy was the result of a social contract among rational human beings who are interested in freedom, but who agree to become socially interdependent, and who then obey laws that they prescribe for themselves (Cheah 2003; Cohen 1986). For Kant, who was inspired by Rousseau, autonomy is located in rational human will, a moral law not ruled by "natural law," in which the rational human self governs the self (Johnson 2014). Autonomy in the liberal tradition centered the rational individual actor and excluded anyone deemed irrational: colonized peoples, women, and certainly nonhuman actors. Yet its conceptualization has offered a means of theorizing freedom under relations of early capitalism in late eighteenth-century Europe and in twentieth-century decolonizing literature in the Third World (Cheah 2003).

According to theorist Pheng Cheah (2003), late eighteenth-century German philosophers, in the wake of Napoleon's rise to power, moved away from theorizing society through mechanistic models as exemplified by Hobbes's *Leviathan*, and instead toward organic models.[38] The claim made by such philosophers as Friedrich Schlegel was that the social body should not be conceived as a hierarchical body ruled by a soul or a mind, subordinate to a government; rather, it should be conceived as interdependent between the parts and whole that make up the body (Cheah 2003; Schlegel 1996).

Cheah's interpretation of autonomy in political theory, of independence through interdependence, often gets obfuscated in everyday politics, especially when considering autonomy in relation to sovereignty and federalism. Writing about Seminole sovereignty in relation to the state of Florida and the federal United States of America, Jessica Cattelino (2008) identified autonomy through a contrast with interdependency. Exemplifying this, white settlers who try to forbid Native American tribes from participating in state and federal politics through political campaign donations make the argument that such sovereign nations should not be able to do so because they should be autonomous, which they interpret as political isolation, a state that Elizabeth Povinelli (2006) would describe as "autological." Seminoles in Florida articulate their political independence—their sovereignty—through legal relations in both tribal and federal law systems, through philanthropy in local settler communities within Florida, and through institutionalized intertribal relations across state lines. For Cattelino, these are expressions of interdependency, not of autonomous isolation. Yet when autonomy is defined through isolation, it will always be impossible to achieve.

The metaphor of the organismic political body took on literal dimensions in the 1970s, when cyberneticists conceived of empirical forms through organismic models of freedom. For instance, cyberneticist Francisco Varela (1979) defined autonomy as a system's ability to distinguish itself from a background, whether that is an orangutan in the forest or staining on a slide under a microscope. Autonomy for Varela is simply "self-law," which he elaborates is a living system's reliance on an internal network of processes and a self-assertion of its identity, especially as it compensates for changes by maintaining its own organization.[39] In this framing, autonomous systems could include single life forms, comprised of multiple systems, as well as abstract assemblages such as corporations or institutions. In this formulation, Varela opposed the human exceptionalism of Kant's argument that autonomy, self-law, was not subject to natural law, but rather was subject to self-imposed laws owing to human rationality. Following Varela, autonomy need not stem from rational thought, but merely from the experience of becoming distinct from a background even as there are porous boundaries that separate interiors from exteriors and facilitate exchanges of inputs and outputs.

Autonomy beyond humanity was salient for Tim Choy (2011) as he joined field biologists studying the genetic autonomy of Hong Kong orchids. These flowers were not just variations on the species of Chinese orchids, but were rather their own species. These studies were contemporaneous with Hong Kong's incorporation into China as an autonomous and distinct entity. This inspired Choy to suggest that rather than thinking through organismic vitality, the metaphor used by Kant and other German-speaking late eighteenth-century philosophers, we could instead think of politics through distinction and specificity.

Autonomy is also generative for Rosemary Collard and colleagues in "A Manifesto for Abundant Futures"; they offer a definition of animal autonomy as "the fullest expression of animal life, including capacity for movement, for social and familial association, and for work and play" (2015: 328). They point out that managerial orientations to wildlife population and the desire to exercise control over them impose humanist colonial and capitalist regimes on their lives, regimes that work to actively diminish their autonomy. They have the autonomy of possessing "their own lookouts, agendas, and needs" and deserve "an abundant future . . . in which other--than-humans have wild lives and live as 'uncolonized others'" (Collard et al. 2015: 328). Yet such a definition of autonomy potentially poses incommensurability: the "fullest expression of animal life" for a subadult male ready to copulate can impinge on the "lookouts, agendas, and needs" of a prepubescent female orangutan on site needing more time to physically grow before even being capable of having procreative sex.

Such different interpretations of autonomy offered by Cheah, Cattelino, Varela, Choy, and Collard et al. suggest that autonomy has to be understood through translation and through the specificity of contexts. Autonomy is different from autology. Its conceptualization has to be understood relationally. Its meaning derives through comparisons, between the English concepts of liberty, imperialist Majapahit courtly concepts of merdeka, and vernacular ideas of bebas and 'ati diri. Its expression is also relational between who feels independent and how they feel it.

'Kak had a pragmatic approach to the problem of what autonomy would mean for her. She needs to work to live within Sarawak's wage economy, and yet she doesn't let this change her own demands for time off to harvest rice or to celebrate Christmas and Gawai, the two holidays of most importance to urban Ibans and other Dayaks of Sarawak. Likewise, compulsory schooling in the federal state of Malaysia means that her children will receive a longer education than she did, which for her had ended at primary school. She has two choices for her children's primary education: either Malay or Chinese. At the time of decolonization, Sarawak's Dayak population fought to keep English as the language of instruction, but they lost to "Bumiputera" politics, which sought to assimilate Sarawak into a Malay-speaking Malaysia dominated by the peninsula (Leigh 1974). By sending her children to Chinese school, she resisted the cultural hegemony a peninsular Malay school would impose on her children, though her children struggle to understand the language of instruction.

For Nadim, the point of autonomy was not to achieve isolated independence, but rather to become responsible. His sense of responsibility is framed by a Muslim sense of environmental ethics. Using one's brain, in Nadim's ethos, would have us extend our imagination to other perspectives, especially the perspectives of semi-wild orangutans like Wani, who likely strive for the autonomy of kebebasan, but are materially prevented from achieving it. This empathy across difference makes sense in the multicultural Sarawak in which Nadim grew up, in which more than thirty distinct ethnic groups have long lived together, in which it was normal in the 1950s to celebrate the end of Ramadan with rice whiskey, which is now considered *haram*.[40]

From Nadim, Layang, and 'Kak, I learned that the sense of autonomy through the translation of bebas is not about performing British ideals of liberalism and freedom, which themselves are structured around the impossibility of autological selves. To be bebas is to have relations of interdependence, to have the possibility of movement under constraint, and to empathetically live with differences, distinctions, and potential risks.[41] Ultimately, fostering independence is an impossible ideal when material conditions foster interdependency: the orangutans depend on the workers for their food and the workers depend on their work with orangutans to buy their food. Yet autonomy here matters for decolonizing Sarawakian politics, because autonomy, following Francisco Varela and Tim Choy, is about acknowledging distinction—distinction of a self from a background and a self. Thinking of autonomy through the concept of bebas compels us to think of this distinction as something to which others must respond (Rutherford 2012). Imagining other perspectives across difference—differences of race, ethnicity, religion, gender, and even species, as Nadim does, is an example of the decolonizing politics of kebebasan. By forming new interdependencies, such efforts to imagine, exist, act, negotiate, and transform might be robust enough to survive imposed material constraints.

Conclusion

According to the official descriptions of wildlife centers, orangutans at these sites are supposed to be trained to achieve autonomy, the kind that is clearly signified in black and white by the phrase *totally wild*. The absence of totally wild orangutans, for visitors at least, proved the existence of the state of being totally wild. Yet being totally wild is a fiction of English terms, intelligible to the few who can speak "white people." The term only occurred in speech when prompted by my questions. The words written on the board were a trace; what was signified by those words was not there.[42]

At these sites, autonomy is the end goal, but it is always deferred and impossible to achieve. Wani's experiences show that the hope of gaining autonomy serves as a means to continue enforcing dependency. Umot and Nini show that independence from one's mother means dependence on human intervention. These orangutans show that the state of being bebas and its pedestrian forms of freedom is conditioned by restraint. Hence, the arrest of autonomy becomes a way to understand the inertia that happens at the wildlife center, just as it helps to understand the state of Sarawak in which independence is indefinitely deferred.

Within the space of the wildlife center, orangutans' lives are managed as both a population and as individual subjects with individual names. The borders of their domain were first drawn by the power of the first White Rajah James Brooke and then further carved up by business interests in sand, concrete, and housing developments associated with the now former chief minister—whether directly through his concrete company or indirectly through his personal associates such as Ting Pek Kiing, whose Global Upline Limited owned the sand mine at Batu Wildlife Center's border before it declared bankruptcy. Within these domains, the orangutans and people who interface each other must negotiate their relationships, as they negotiate between the abstractions of autonomy and constraint. Who must acquiesce to whose wishes? How can one move through the site, literally when trees acting as corridors are removed to build toilets, and figuratively when one's opinion is never solicited, when what should be done is impossible to do?

Through the movement of vision between scales of time and space, between the peripheral province and the central nation, the Raj, and the wildlife center, we see how the concepts of freedom and autonomy captured in the term *bebas* connect the lives of wildlife and people in Sarawak. Within the domain of the wildlife center, we see that those who appear to be free are really not, that the state of being bebas has its own conditions, of what Wani experiences when she has to return for food, what Nadim saw as fear, what Layang saw as forests transformed into oil palm, and what 'Kak saw in her underpaid wages. Their worlds are shaped by an arrested autonomy. Within the realm of the private colony of the White Rajahs, we see that the state of being bebas and *senang*, moving with "freedom and ease," is the privilege of rulers. Within the realm of the contemporary nation-state and province, we see that the ability to determine what can become sand, concrete, dams, and plantations is an exercise of license and unrestrained freedom. It is in the semi-wild world of the wildlife center that we see how the state of being free and having license is exceptional and denied to most.

Chapter 6

HOSPICE FOR A DYING SPECIES

With its tall concrete walls, wrought-iron bars, damp concrete, and rusted--over steel cages, Lundu Wildlife Center did not look like the site of unrestrained freedom evoked by the adjective *bebas* to anyone who saw it. It looked, smelled, and felt like a zoo, perhaps even one that had failed accreditation.[1] Layang and other animal keepers knew this and resented it. When he worked for more than a decade at Lundu's parent site, the free-range site Batu Wildlife Center, he and his colleagues would correct anyone within earshot who would call that site a zoo: "*Bukan zoo*" (this is not a zoo). Thus, when Ching and Ti were held in captivity throughout their pregnancies at Lundu, Layang was frustrated. He felt that their captive conditions would surely lead to fatality, as it had for the last mother–infant dyad. Both had died within two weeks of the infant's birth just a few years before.[2] That infant's bones still sat on a shelf in a plastic bag in the nursery, which had been the veterinarian's laboratory more than a decade ago when the site was a government entity and had a full-time veterinarian. Layang's frustration concerned not only his conflict with the administrators, who ignored his insistence that semi-wild conditions should be met, but also what volunteers would think. With the volunteers in mind, he once said to me: "They think the zoo in their country is better than this. And this is supposed to be a rehabilitation center."[3]

Few entertained such pessimism; instead, they actively sought examples of how the wildlife center worked toward the goal of orangutans' eventual independence. Early in my fieldwork, when I was living alone in staff quarters, I hosted a dinner with two volunteers in their twenties. Over instant noodles with seaweed, we got into a conversation about zoos. The young woman with facial piercings spoke with an elite accent, having grown up in Kent in the south of England, and was taking a break from her undergraduate studies in anthropology. She was vehemently against the idea of zoos. Without citing art critic John Berger (1980) by name, she referenced points that were older than she was by half a decade: humans had no right to gawk at animals, zoo settings were unnatural, and orangutans should be in their natural environment (Uddin 2015). The young man was one of the very few single men who participated as a volunteer. He spoke with a working-class accent that revealed his upbringing in East London. Having trained as an accountant, and living with his parents, he had a career in sufficient demand and a low cost of living that he could afford to take long trips abroad between jobs. He asked her a question: "Are you against the idea of nursing homes?"

The young woman refused the connection and said, "an old person has a choice in going to the nursing home." The young man replied that was only sometimes the case: his grandmother was forced to go to a nursing home because she was deemed a danger to herself. His parents didn't have the money to get a caretaker, so she's there, against her will.

The volunteer's description of the nursing home offers a provocative comparison not only to zoos, but also to the care of wildlife at the orangutan rehabilitation center. Such sites of care are ultimately commercial enterprises. Even if they are not profitable, they need to stay afloat if they are to continue offering care in the face of death—whether at the level of the individual or the species. Thinking of these institutions in conversation with each other shows that biopolitics—the management of life—has its flip side of necropolitics, management through death and the threat of dying (Foucault 1978; Haritaworn et al. 2014; Mbembe 2001; Wolfe 2003; Wright 2011).

Following the volunteer's analogy, I am led to think of the wildlife center as a kind of nursing home, a place of care at the very end of one's life. The nursing home, and its more dignified counterpart, the assisted living center, references the institutionalized and supervised confinement of those unable to survive alone, who are given constrained conditions in which to live through compromised independence, and whose caregivers support their living while also actively anticipating their imminent demise. An institution related to the nursing home is the hospice. Hospice is the place of care at the threshold between life and death in which resources of money and care are exchanged (Russ 2005). Hospice is also an apt analogy when considering the care of cancer patients, for whom medical treatment is either toward a cure and survival or is palliative and looking toward death (Jain 2013). As Tom had said to me in our very first conversation, commercial volunteerism in Sarawak does not promise to end extinction; it promises small acts of help. Orangutan rehabilitation can only offer species-level palliative care to the end, without expectation of the species' survival. Layang, Lin, and their coworkers might scoff at the comparison, and yet their workplace undeniably offered institutionalized care: the same supplier who sourced food for the wildlife center also sourced food for nursing homes.[4] In this chapter, I use hospice as the term to reference end-of-life care at an institution. Here we see that the wildlife center is a hospice operating at a scale that collapses individual with species.

The wildlife center has had more than its share of being a threshold of life and death on the scale of individuals. The orangutan mother and infant that died within two weeks of the infant's birth (this chapter), the small wildcat that died of rickets (chapter 3), and the crocodile that was euthanized and then dissected on site near the orangutan nursery (chapter 4) are just three examples I have shared in this ethnography.[5] Sometimes displaced animals come to the center so ill and mistreated that death is imminent and seemingly welcomed.[6] Care in such cases was palliative, experienced intersubjectively between the dying and those around them as their final moments passed.

I suspect that orangutans after extinction in the wild will be forced to live in either rehabilitation centers or zoos, both of which foster end-of-life care for dying species. Rehabilitation centers, with open space and possibilities for free-range mobility, are contrasted with zoos, and yet both the zoo and the rehabilitation center fulfill the same purpose: they facilitate human encounters with species facing the threat of extinction that would otherwise be too difficult to see in the wild.

This chapter considers the wildlife center as a commercial hospice. It describes the disturbing ambiguity of the term *wildlife center* as experienced by commercial volunteers and wildlife center workers. The distinction between zoo and rehabilitation center, which was crucially important to workers like Layang and volunteers like my Kentish dinner guest, proved to be tenuous. A series of confrontations between orangutans and care workers on site broke the distinction between rehabilitation center and zoo. These physical confrontations highlighted how sacrifice in the work of care involved intersections of gender, race, and class that exceeded human contexts. This chapter engages the question of who bears the burden of end-of-life care when the scale of life and death conflates individuals with species.

Bearing the Cost of Care at the Wildlife Center

What was the wildlife center? Was it a rehabilitation center or was it a zoo? This was not merely an issue of semantics. On a conceptual level, zoos for Layang and his coworkers perpetuated and even aspired to relations of unequal and colonial domination. Layang's reasoning that zoos were colonial spaces was not because they forced orangutans into unnatural settings, as my Kentish dinner guest felt. Nor was it about the imposition of Western scientific ideas of species distinction and zoos' representation of colonial domination, as many scholars have argued (Rothfels 2002; Uddin 2015). Rather, zoos in Layang's opinion were colonial because of the attempt to tame animals. He used the example of Singapore Zoo's famous Amek the Orangutan to illustrate the point to me. She was trained to follow commands to sit still near people while they ate luxurious meals in her company. Layang instead espoused the position that if this place was a site of orangutan rehabilitation, it was about the paradox of tough love, in which specific people like himself maintained relations with orangutans so that the latter would be generally afraid of people. The kind of relations at stake in rehabilitation was very different from domination: it was mutual but unequal vulnerability. The shared risk and uncertainty of working together without barriers and without the pretenses of subjugation and domination exemplified the decolonizing principles of being bebas. On a practical level, the difference between a zoo and a rehabilitation center lay in the work of care. Like all forms of care, the work of care at Lundu Wildlife Center pays caregivers, but it also requires caregivers' payment of sacrifice in return. This raises the question of who sacrifices and how much (Govindrajan 2015).

Not all workers at the center shared Layang and the volunteer's conviction that the zoo was so different from the rehabilitation center. Lin was the junior officer and second in command of the center, just under the person in charge in the company flowchart. She spoke English with me, which she had learned as a fifth language in secondary school and which she honed through downloaded episodes of *Gossip Girl*, an American television series about elite young New Yorkers.[7] Her own background was quite different: her parents ran a simple Foochow Chinese food stall at a hawker center in Sibu, Sarawak's second-largest city, famed as a rough port city run by gangsters. She was the first in her family to go to university. She did not pick her field of study of forestry: it was given to her after her national college entrance exam, as is the case for all students in Malaysia's public universities. When we first met in October 2008, she was still new at her job, having been there only a few months. Her many duties included writing reports, managing the center's programs, accompanying overnight and weeklong jungle skills training sessions, and caring for infant orangutans a few nights a week. Such overnight work was given to her and her fellow officer Cindy, who was also a young college-educated woman. While Lin could see herself as upwardly mobile, Cindy was on a downward social trajectory. Her Chinese father was a lawyer in the city of Kuching and her Iban mother was a nurse. When she finished secondary school the same year as the Asian financial crisis of 1997, her plans to study medicine in Australia were dashed with the devaluation of Southeast Asian currencies. She, also without a choice as to her field of study, pursued biotechnology at the local university.

For Lin, the difference between the wildlife center and wild habitats was greater than the difference between wildlife center and zoo. She conveyed this in an interview. I had asked Lin to reflect on the idea that orangutan rehabilitation was about "tough love," which was the common idiom workers used to describe the work of care at the rehabilitation center. These acts of tough love included hitting orangutans, turning away from them, and repelling their close contact with eucalyptus oil rubbed onto wrists (see chapter 2). These acts were meant to give them the possibility of becoming rehabilitant orangutans instead of captive orangutans kept for breeding. Such acts should be construed as acts of care, even if they are violent (van Dooren 2014). The hitting of young orangutans as acts of care and tough love challenges our assumption that care is necessarily tender (de la Bellacasa 2012; Martin et al. 2015; Murphy 2015; Parreñas 2012; Schrader 2015; Stoler 2002). The work of care at orangutan rehabilitation centers and zoos challenges our gendered assumptions that the giving of care is maternal.

Lin interpreted my question about her thoughts on tough love not to be about performing toughness with her orangutan charges, but rather to be about the toughness she herself experienced on the job. As she saw it, work at the zoo and the wildlife center was similar because both involved intersubjective relations with her charges since they share a state of captivity. The bigger difference for her was the distinction between wildlife center and wild circumstances. Primatologists and conservation biologists in the field maintain great distances between themselves and their arboreal orangutan subjects. Actual orangutans may not even necessarily be part of these studies, since their traces in the form of vacant nests act as their proxies in some methodologies. Lin suggests that the intersubjective relations across species at rehabilitation centers was always gendered:

> Rehab is not an easy. . . . You just ready yourself to be injured, anytime. . . . The risk is always there. . . . The most common risk is you have to be ready to be bitten by them, just like Cindy. . . . *If you as a lady join this conservation you really have to sacrifice a lot. You have to sacrifice. I would not only say tough, but I would include sacrifice also.* Yeah. And because you are . . . having a relation with them, handling them, you have to be ready if one day, this animal is going to transfer any disease to you *or* you transfer any disease to them. And probably you bring that disease *to your family* [her emphasis]. So that is the things [*sic*] you have to, the concerns you have to think about when you want to join this life. That is what I understand on the tough part of that rehab. At least, *if you are on the wild orangutan project, you're just tracking, finding a nest. You will never—not say never, you will seldom meet with the wild orangutan. Even though you meet with them, you are having a very far distance from them. You don't have any contact with them. But this orangutan in captivity, along with rehab, you will have to keep touching them, playing with them, just asking them, "how are you," something like that, so it's totally different. Totally different. Yeah, of course, but again, if you don't love your job, the toughness part, the sacrifice part, you won't do it.*

The tough part of Lin's job as Lin saw it was not rejecting the orangutans' demands for tenderness or participation in intersubjective relations. Rather, the tough part was the personal sacrifice she takes on by doing this work specifically as a young woman expected and wanting to get married and have children. This work demands a sacrifice in which she potentially loses her femininity by gaining unfeminine scars from leech or insect bites and a future intimacy with her own children, should she have them.

Femininity for her, as for her generation and her parents,' meant retaining unscarred beauty. Scars are considered ugly for urban and rural women alike. In cities, scars mark bodies that do not have access to mosquito-free air conditioning, or at the least simple electric fans that help deter mosquito flight paths to human flesh. In villages like the Iban longhouse outside Lundu Wildlife in which I lived for seven months, women gossip about other women's ugliness by talking about the scars on their legs from insect bites. Femininity for Lin was embodied through motherhood and the gendered expectations of maternal care. Should Lin have children of her own, she would feel prevented from going on weeklong rehabilitation trainings in the jungle. She already predicts a future for herself in which she has a "second shift" as a mother and primary caretaker of her children (Hochschild 2003). In such a future, caring for orangutans would compete against caring for children.

Acts of care and professions that provide care are gendered as feminine. Hospice, as a form of care work, is often seen as femininely gendered. The contemporary concept of hospice arose in the late 1960s in England and the United States. The point of hospice was to offer people a way to die under less clinical circumstances than hospitals and in more intimate and intersubjective social contexts. Hospice was a form of deinstitutionalizing medical care in the First World/Global North at a moment when Western medicine defined the pinnacle of modernity and development for the Global South (Mor and Allen 2006). In the decades since, hospice has become institutionalized and professionalized.

When thinking of the institutionalization of care, hospice bears functions analogous to those of the wildlife center. It is not contained in the clinical space of the hospital under the control of formally recognized doctors, whether medical or veterinary. Nor is it limited to the home and dependent on unpaid labor. It is instead a third option, one that fosters a special intimacy in the face of death.

Returning to Lin's words, we see that the intimacy of co-presence with apes meant the possibility of sharing illnesses. The gravity with which she spoke of "disease" and zoonotic transfer made that especially clear. Lin seemed convinced that the only possible compensation for the vulnerability entailed in this job was personal love for the work. I was left wondering if her conviction was meant to also convince herself that this vulnerability was worth it, especially when the present conditions of her job already posed for her problems in how she foresaw the future.

The cost of receiving a bite was not only gendered in respect to the impact of disfiguring scar tissue or familial obligations. The moments preceding a bite were also construed as gendered. The orangutan Ching had a reputation for biting women. I was repeatedly warned that she didn't like "local women" in particular. Lin, who was Chinese, her coworker Cindy, who was mixed Iban and Chinese, and I, a Filipina American who was often mistaken to be Iban or Chinese, were all likely targets. The men who worked under Cindy and Lin repeatedly joked that Apai Len, a grandfather who worked on the site and who had feminine mannerisms, was also a likely target.

Cindy received a bite from an orangutan during a jungle skills trip—not from Ching, but from the ten-year-old female Lisbet. Cindy was embroiled in a confusing encounter involving four orangutans; in the commotion, Lisbet bit her three times on her hand. Part of the job during these and any sessions in "the jungle" of the wildlife center was to discourage human contact. Her male coworkers preferred doing so with violence and the threat of violence. Cindy preferred using her body to convey a sense of rejection. While this method worked well enough during brief two-hour sessions, it was untenable for the longer visits at the ranger station in the jungle, situated an hourlong uphill trek from the center.[8]

Cindy reflected on the incident with me weeks later, when she reported back to work caring for the infant Lingga overnight in the vet clinic. She explained that the orangutan Ching, who was twenty years old at the time, had raided the workers' supplies while they were in another part of the forest with ten-year-old Lisbet, four-year-old Gas, and three-year-old Lee. Before her coworker Layang trekked downhill to get more food for them, he advised her to boil a pot of water and throw it at Ching if she were to again enter the interior of the ranger station. The orangutan Ching did enter again and Cindy's colleague Apai Len threw the pot of water at the orangutan. By then the water was cold and a chase ensued between twenty--year-old Ching, Cindy, and the two older men whom she managed. In the commotion, ten-year-old Lisbet grabbed hold of Cindy and bit her three times on her hand. It only ended when Apai Julai hit Ching with a wooden plank and she retreated into the forest.

This led to the expectation that Ching would exact revenge, which came weeks later when she joined Apai Julai on a platform during a heavy rainfall, sat beside him, and then bit him. Apai Julai had described this to me using the term *sakit hati*, a resentment and injustice that pains one's innards, as a way to explain what he thought Ching felt.

This led to Lin taking preventive action: As soon as she saw Apai Julai, she photographed his bloody forehead and immediately contacted the regional director, going over her immediate boss, Syed. The regional director issued an order to put all the orangutans back into captivity. The workers under Lin had to comply because their employment hinged on their fulfillment of this duty.

Workers like Layang, the volunteer company project affiliates Tom and Ben, and Syed, the person in charge of the park, were all angry at Lin for initiating this action. One worker expressed his disapproval by writing on the cover of the daily nutrition diary, crossing out the words Wildlife Center and replacing them with "zoo."

The figure of the zoo was deeply antithetical to Layang and the decolonizing possibilities that are open when orangutans are bebas or autonomous at the wildlife center. The physical autonomy in which they can self-locomote in a variety of ways—adapted to climbing through the tree canopy over millennia and adapted to walking within their individual lifetimes—makes it patently clear that these individuals are not deathly ill and do not need hospice care. However, the inability to live outside the wildlife center, because of habitat destruction and the inability to live peaceably under human domination in a limited definition of peace, means that they have no choice but to live in such an institution. The wildlife center, in Layang's ideal vision, would recognize individual abilities while recognizing species-level impossibilities in the present and projected future.

Tom, the British volunteer company project manager, supported Layang's vision. However, Lin, the Chinese Malaysian junior officer, did not, and she had good reason not to: the orangutans' autonomy would mean that Lin and Cindy would be more vulnerable to violence than others on site.

Cindy, Lin's colleague at the same level within the company hierarchy, didn't understand why everyone else was so vehemently against the idea of zoos. Cindy pointed out to me that animals look healthy at zoos and that zoos allow animals to be closer to people who have difficulty jungle trekking. While Layang hated Singapore Zoo, Cindy loved it: she loved its white tigers and polar bear. And when she started working at Lundu Wildlife in 2010, she did not know that a "wildlife center" was technically different from a zoo. Though everyone corrected her use of terms, she thought it was strange that all her colleagues were unclear about the definition. She said, "These people keep telling me a wildlife center is basically a rescue center, but not really also. I don't know!" She suggested I ask the general manager of the corporation. Such a question could sound either naïve or critical when posed to the general manager, one of the highest figures on the company organizational chart, just below the board and chief executive officer. Coming from me, the question could be interpreted as an insult toward the institution hosting me. I chose not to ask him.

Tom, who felt that the support of his volunteer company rested solely on Lundu's orangutan rehabilitation functions, was especially frustrated. He then worked to force the Forestry Corporation to resume rehabilitation efforts. He did so by orchestrating sick leave among all the workers whose salaries his company subsidized. In effect, it was a strike. Syed convinced Tom to stop his protest and to resume his support because it would likely end badly, with the Forestry Corporation canceling their memorandum of understanding, at which the commercial volunteer company would have to immediately leave Sarawak and never be able to return. Syed was nonetheless very angry, because Lin had undermined his authority. The problem for him lay in what he felt was Lin's refusal to share the quintessential form of relation at the wildlife center. That relation was ultimately about mutual, although unequal, vulnerability via bodily sacrifice.

Syed reflected on this in our last interview. He refused to let me record, but he allowed me to jot notes:

> "How can this be a center for rehab, when you want all the orangs in a cage?" Syed said exasperated. "Rehab is hard. . . . Just because you study, you think you can be white collar. It's not like that. Listen, if you think 'I'm in power,' it's going to fail you. I'm frustrated. I cannot say anything. My authority is over-righted [*sic*: overridden or

overruled]. The target, mission, endeavor of rehab, of conservation is slow. You have to swim in hot water. Rehab is not just about the little animals. That animal is growing up. What are you going to do? Put in a cage again? Conservation takes sincere objective, not ego. The problem: some do not want to accept the reality of danger. You have to sacrifice and have a hard time. Your question is the hot topic. You're a professional observer. I think you know."

While Lin had stressed that the work of care for a species facing extinction was gendered, Syed stressed that the work of care was about class, that this job was not about writing reports on laptops inside air-conditioned offices, but rather sweaty and dangerous work outside in which, as Lin said, "risk is always there." Both Lin and Syed, of course, are right, that gender and class shape how the work of care gets done, especially when all the workers understood that Ching had a propensity to attack women. If sacrifice and mutual vulnerability are at the heart of the matter in the work of care here, we have to keep in mind who sacrifices more in these states of vulnerability.

In this respect, hospices offer another point of comparison. Those being cared for are vulnerable to workers, while workers are vulnerable to workplace injury. The banal risk of injuring one's back when lifting a person who cannot lift herself, or everyday verbal abuse dispensed by elders resentful of their dependency on others is not as shocking as the risk experienced by elder care charges vulnerable to abuse by their caregivers. Such violence is frequent enough to have an academic journal dedicated to the subject (*Journal of Elder Abuse and Neglect*, published since 1988). In these relations of care, each is differently vulnerable to the other, and yet some risk more than others.

Having stressed the lack of ego necessary in performing the work of care, Syed dismissed any feeling of heroism one might gain from an altruistic sacrificing of one's safety for the sake of orangutans' free-range freedom. Living together with others across species differences, sharing and inhabiting the same space, inherently meant subjecting oneself to hard work and pain without heroism. This was the everyday work of care here—whether wildlife center, orangutan rehabilitation center, or zoo: it nevertheless acted as a hospice for a dying species that made differently difficult demands upon its caregivers.

While she was recovering from the infected wound that resulted from the bite, Cindy's parents begged her to quit. She declined. We spoke about it one evening after she had succeeded in putting the infant orangutan to sleep. She was going to stay the night in the quarantine, watching over the orangutan. Even as her hand was still bandaged up, even as she anticipated a sleep-deprived night on a plastic folding beach chaise, she could tell me with all earnestness that she saw her job as a privilege.

Capitalizing on Future Death

Work at the wildlife center demanded sacrifice and a willingness to experience physical vulnerability. Yet sacrifice could be seen as a privilege, both by commercial volunteers paying thousands of dollars for the experience and by employees like Cindy. Knowing all this was the likely impetus for the Forestry Corporation to begin exploring other possible funding streams and new consumers. The corporation's capital could not come from investment banking in the form of shares: they were not a public entity with shares available for trade in the stock market—at least not yet. The futures worried about by those in the Forestry Corporation's headquarters were not the financial instruments of derivatives, which allow for the buying and selling of projected performances (Ho 2009; Zaloom 2006). Their concern for the future was about the source of their future customers.

The way senior managers of the Forestry Corporation conceived of who their clientele was or ought to be suggested limits to the analogy of the wildlife center as a hospice or nursing home. In a hospice, care recipients and their families are the clients. In a Forestry Corporation wildlife center, orangutans are not clients; paying human visitors are clients. The orangutans were assets when the Forestry Corporation could commodify encounters with them. A few executives piloted revenue-generating programs independent of the British volunteer company. They were hoping to gain clientele from other Asian countries and single-day VIP corporate social responsibility visitors for day trips that involved working with the orangutans on site at Lundu Wildlife Center.

This was a different kind of transaction from Dutch retirement homes that offer subsidized housing for college students in exchange for their emotional and affective labor of socializing with retirees (Hochschild 1983; Muehlebach 2011; Young 2015). The mutual benefit of the transaction between day-tripping clients and orangutans is simply lacking, especially when considering Nadim's observation over years of working at Batu Wildlife Center that visitors "never think what the orangutan are thinking." Day-tripping clients would be unable to build a social relationship with their temporary charges. While all social relations are fleeting, this particular interspecies interface would especially be short-lived.

The idea of greenwashing is a relevant one here. Following Greenpeace activists (Greer and Bruno 1996) who coined the term while working in Malaysia, scholar Sharon Beder (2002) defined it as environmentalist public relations rhetoric that corporations and other entities engage in as a way to cover up their environmentally detrimental practices. The charge of greenwashing suggests duplicity—cynically using empty words to mask one's questionable actions. Yet the Forestry Corporation sought something different. Greenwashing is often about corporate marketing strategy, the identity of the company that becomes an asset propping up the reputation of the corporation. The Forestry Corporation sought to use such efforts not for its own marketing, nor the marketing of other companies using corporate social responsibility initiatives. They instead sought to use the experience of co-presence with orangutans as an asset itself. Such embodied experiences were more valuable than the reputation any greenwashing effort could gain. While any experience is fleeting, the particular experience they wanted to capitalize was already very scarce and was solely theirs as agents of the state government responsible for wildlife. Hence, they could potentially make serious profits from the future demise of wild orangutans.

In my last conversation with Syed, I mentioned that the regional director had told me that the Orangutan Conservation Fund was intended to support the wildlife centers without the help of state funding. Syed was flabbergasted: "The Forestry Corporation's a mess. Conservation without government assistance? Bullshit!" It was clear that he felt that protecting wildlife could never be a private industry, that profitability or even "financial independence" was impossible. The survival of the wildlife center being able to offer care for a species facing extinction was contingent on continued public funding. Without such state assistance, what would the future hold for a privately run corporation charged with the care of the state's endangered species?

The twenty-nine-page law of 1995 that created the Forestry Corporation shows how the corporation itself, officially working as an "agent of the government," both operated at the thin edge between private and public sectors and was also intended to act as a financial instrument in itself. Who the beneficiaries of that financial instrumentalization were was not explicitly stated. Yet the apparatus of finance left much to the discretion of the board, five to eight persons appointed by Sarawak's minister responsible for the state's resource planning, who, until recently, happened to also be the CEO of the Forestry Corporation.

The Forestry Corporation was unlike typical corporations. It was not a publicly traded company and thus did not have shares or shareholders (Welker 2014). Neither was it like early twentieth-century American businesses that grew to become corporations before the rise of the stock exchange, in which gains were made through retained earnings, personal family wealth, or commercial bank loans (Ho 2009). The Forestry Corporation technically consisted of two companies: the Forestry Corporation that received government contracts and the Forestry Corporation private limited company that contracted staff to fulfill those government contracts (Chan 2008). Through this mechanism, the state of Sarawak was able to carry out the duties of the Forestry Department with significantly less cost to the government (Chan 2008).[9]

Syed, as Lundu Wildlife's person in charge, experienced the Forestry Corporation through its special relationship with the state government, and specifically the government's payment for services performed by the corporation on behalf of the state government. The Forestry Corporation received and distributed state and federal government grants via the Forestry Department. Yet other potential funding streams specified in the law opened up possibilities of financializing the wildlife center. This included the potential to monetize the Forestry Corporation's assets through leases or even liquidating assets through disposal. This also included the power to invest, as broadly defined, whether in stocks, bonds, derivatives trading, or real estate. Additionally, it allowed the Forestry Corporation to take on debt. The law reveals how the power of the Forestry Corporation lay not so much on its task to "manage and administer, on behalf of the [state] Government, all forest reserves, protected forests, national parks, nature reserves, wild life sanctuaries, and areas reserved for forestry research, recreation and conservation," but rather on its utility as a financial instrument. The Forestry Corporation's investments are managed as the board sees fit:

they are permitted to spend state monies that would have gone to a state agency but are instead going to a private corporation.[10]

The schemes that demanded participation from officers like Lin and subcontracted workers like Layang excluded the British commercial volunteer company. They required that people like Layang and Lin reconceive of their work of care not simply as caregivers for the needs of orangutans on the brink of extinction. Instead, they had to conceive of themselves as entrepreneurs who sell their services to the Forestry Corporation and who must cultivate feasible conditions for the Forestry Corporation's clients. These clients were not orangutans, but tourists and travelers keen on seeing orangutans in the flesh before they die off.

Layang had long felt that the corporation and its entrepreneurial vision was bound to fail: "You can't corporatize conservation. Sabah, *Semenanjung* [peninsular Malaysia] doesn't corporatize. They're under PERHILITAN [Jabatan Perlindungan Hidupan Liar dan Taman Negara, the Department for the Protection of Wildlife and National Parks]. If you corporatize, you need a product. What's your product? Batu [Wildlife Center] has orangutan. What's Lundu's product? People come here and they see Ching in a cage" (November 12, 2009).

Layang's question was rhetorical, yet the answer stopped dead in the cold bars enclosing Ching and the others. Orangutans are not in themselves products made through the labor of their caregivers who try to cultivate wild orangutan behaviors among orangutans acculturated to human behavior. Rather, the sensory experience generated between different bodies in the space of the wildlife center is the product. This objectification occurs in ways that John Berger (1980) likely could never have imagined, and in ways that my two dinner guests probably felt but couldn't describe.

Palliative Care without Comfort

An accident had killed the orangutan Taib at Batu Wildlife Center. When he extended his arm to grab a rope, he likely had no idea that he was reaching for a live and uncoated electric wire. The whiteboard bearing the names of all orangutans on the site did not have his name written on it. He was named after the chief minister at the time, who was serving the longest term of any chief minister in any of Malaysia's states and whose decades in power were marred by accusations of corruption, cronyism, and deforestation for personal profit. The embarrassment of Taib the orangutan's demise was not just about the negligent circumstances of his death, but what such circumstances revealed about the priorities of the state of Sarawak under Taib the chief minister. The wires have since been coated, but the plan to bury the wires underground so that no such mistake could happen again has been "put in view" for more than a decade. Nadim explained the concept of "put in view" by mimicking his hands as ledgers. The plan to bury wires is on a list of proposed updates that never seem to get funded. He closed his hands, silently showing how it is put out of view.

A different kind of accident killed Bullet. Bullet earned his name when he survived a farmer's bullet through his face, with its large flange. He was shot with a tranquilizer. The first injection, delivered by a blowgun, did not seem to work. He was still climbing. The second shot got him, but did him in. His heart stopped. The Forestry Department decided to preserve his corpse and put him on display at Lundu Wildlife Center at the office entrance. His taxidermied body was moved to storage when the building was renovated in the late 2000s. After all, tourists were there to see live exhibits, not dead ones.

Arthur also died from a human misunderstanding. He was one of the last orangutans to be hand-reared by Barbara Harrisson. Arthur and Cynthia were part of the first-ever orangutan rehabilitation effort, which ran between 1962 and 1964 at Bako National Park, in the midst of Sarawak and Sabah's official decolonization and federation with Malaysia in 1963. Bako National Park was untenable for rehabilitation purposes. It offered less than 10 km^2 of forest habitable for orangutans, and the orangutans frightened beach-going tourists.[11]

Arthur and Cynthia were sent to Sabah, on the northeast of Borneo, from whence they came and where an orangutan rehabilitation center was starting to open under the guidance of the game warden Gananath Stanley de Silva. He was Ceylonese, and he took the position during the colonial era of British North Borneo and retained it for twenty-five years. The forests at the time were in his words "being raped," and the revenues made the nearby port city of Sandakan a city of millionaires in the 1970s, which would be hard to imagine for any early twenty-first-century visitor.[12] The job of orangutan rehabilitation attracted only those needing a job, and they did not necessarily have an interest in the demanding work of rehabilitat-

FIGURE 6.1 Barbara Harrisson with Arthur and Cynthia in Bako National Park. Note the stains from leech bites on her socks. Image reprinted with permission from the Sarawak Museum.

ing orangutans. One day when Barbara Harrisson was running errands in Sandakan, de Silva's worker shot and killed Arthur.

As I sat in Barbara Harrisson's living room in the Netherlands in 2006 and asked her to recall the aftermath of events nearly four decades later, her raised voice expressed her continued agitation and anger: "De Silva could not punish the civil servant. He was Ceylonese, just like de Silva. De Silva could not punish his own countryman on the command of a British colonial."

In de Silva's recollection, the man who pulled the trigger wasn't the Ceylonese worker, but the Malay. Both workers told de Silva that Arthur was "running amok," pushed over the water tank, and cornered the two men; one of them shot him.[13] De Silva, recounting this to me in 2007 on a picnic bench outside a shopping center in a multicultural suburb of Toronto, doubted the story his workers told him. There was no trace of water on the ground. De Silva figured that the true story was that Arthur had entered the hut in which the two men stayed; out of fright, one of the men fatally shot him.

These orangutans were otherwise healthy, but met their untimely demise because of human-caused accidents. These stories occurred in the span of time following official decolonization, when orangutan rehabilitation was an effort carried out by the postcolonial state. Much like a fatal overdose of anesthesia, death by excess tranquilizer is more common than anyone would like to think. Yet when considered along with Taib's death by electric shock and Arthur's fatal gunshot wound, such deaths together convey indifference. There is no indication of a moral obligation to orangutans in this indifference. It is indifferent to Nadim's pious and Islamic sense responsibility toward his charges. Likewise, such negligent treatment of orangutans is void of any sense of Iban obligations to potential ancestors. The secular bureaucracy of the postcolonial state agencies responsible for orangutan rehabilitation efforts means that any potential lessons from these stories are simply put in view.

If the postcolonial intervention for orangutan rehabilitation has been to record these fatal stories for the Forestry Department archives, which are sealed from the public but accessible when people share their memories of working at these sites, then what is a decolonizing intervention? Would it be the story of Cynthia? Cynthia was Arthur's peer. De Silva reckoned the two orangutans both felt Barbara Harrisson's absence when she transferred them to Sabah. Sometime after Cynthia lost her companion to the fatal shooting, she was seen examining a clay mound with a hole in it. She was then seen with one of her hands swollen and blue. She was later seen with her entire arm swollen. She then died, most likely because a cobra inhabited the hole in the clay mound. Her veins were probably filled with venom. Nobody intervened because nothing could be done. This is palliative care that offers no comfort.

Conclusion

When I left Sarawak in July 2010, two subadult twelve- and fourteen-year--old males Aqil and Ahmed at the free-range site Batu Wildlife Center copulated with eight-year-old Mia in what onlookers interpreted as both violent and forced (see chapter 3). Her unusually slow movements were a cause of concern for workers Nadim and Boboy. Boboy had called Tom, who then called the veterinarian. Peter, the person in charge at Batu, reiterated his strong conviction that all copulation, including copulation with a prepubescent eight-year-old that could not possibly be reproductive, was nonetheless natural. In the end, it was decided that nothing could be done, that she would likely get better on her own, that any intervention involving an attempt to forcibly catch her or sedate her while she was high in the trees so that she could either be examined or taken to another sanctuary would be more harmful than helpful. Everyone at Batu Wildlife Center agreed: an intervention would likely cause more harm than good.

The decision not to intervene is what makes the wildlife center remarkably different from the nursing home evoked by the volunteer who thought about his grandmother, held at an elder care facility against her will. This lack of intervention is akin to the palliative care of hospice, of knowing nothing can be done to heal the underlying reasons for which the body being cared for is sent to an institution and held in captivity until death. Encounters with orangutans kept at these sites serve as greetings to individual orangutans and goodbyes to their species.

Conclusion

LIVING AND DYING TOGETHER

Ice ages and rapid climate changes during the Pleistocene induced the evolution of Bornean orangutans, enabling them to live off tree bark in times of scarcity. This adaptation is a testament to their resilience. Knowing this might lead some to optimistically conclude that orangutans could last through the Anthropocene, too. But it's unclear how they might actually manage to survive through catastrophes that have become normal, such as the Southeast Asian haze of 1997, 2006, and 2015. I imagine that the orangutans who can breathe and withstand smog will have the best chances to thrive. The elemental power of the haze serves a reminder that even when contact is not directly embodied and felt at the surface of the skin, future orangutans will be forced to live with humans and their traces.[1] Under these circumstances, living together also means slowly dying together.

I suspect that orangutan survival will rely on institutionalization, whether in wildlife centers or in zoos. This future seems especially likely when agricultural industries in Borneo continue to convert biodiverse forests into mono-crop plantations. As wild orangutans face displacement and increasingly end up in rehabilitation centers, they are bound to experience constant social interaction (Husson et al. 2016). The solitude that typifies their species will be elusive, if not impossible to find, as they are forced into closer contact with other apes as well as humans.

When I last saw Gas and Lee at Lundu Wildlife Center, Gas had been the larger of the two. I could not have fathomed a time, just six years later, when Lee would outgrow Gas in size and strength. Eight-year-old Lee sired a baby with Gas, who gave birth when she was about ten, four to five years younger than a typical first-time orangutan mother. Gas stopped feeding and caring for her infant two weeks after giving birth. That infant was taken from Gas to be raised by keepers. Ching remains in a cage while Sibu and James have developed cheek flanges. Ti, however, is bebas.

All the officers at Lundu Wildlife had changed postings. Lin had moved to headquarters, while Cindy works at another park. Syed's generation, which includes Nadim, gained retirement pensions. The Forestry Department had reassumed many of the duties previously outsourced to the Forestry Corporation. Currently the Forestry Corporation's primary function is tourism, which includes managing both of its orangutan rehabilitation sites, Lundu and Batu.

At Batu Wildlife, the orangutans that were marked as wild in 2010 no longer appear on the public list. Morni, Umot, Wani, and her child Ros, born in 2005, have not been seen in years, but they are still factored into the official population figure at Batu Wildlife Center. In 2016 that figure was twenty-eight: new births offset official deaths.

Norma, Deh, and Jeffrey died. Norma suddenly stopped appearing at the feeding platform when she had hitherto been a regular. One day her baby who was less than a year old came out from the forest, crying as he struggled to climb. That baby was sent to Lundu Wildlife Center, where he was released after 'Kak cared for him. He was officially designated as abandoned.

Nadim and his coworkers are convinced that Lucas killed Deh after he spent weeks chasing her whenever he saw her. When they heard a strange noise from Lucas in the forest, they tried to approach, but felt it was too dangerous for them to get close. They could see from a distance that Lucas was near a pile of leafy tree branches. Once Lucas left the area three days later, the men found Deh's corpse beneath the pile, with wounds cut deep into her neck. Her live infant, still clutching her mother's body, suckled in vain. In the moments before workers dealt with the body, Nini, her sister, scooped up the orphan. The baby survived—or at least she was still alive at the age of seven in 2016.

Deh's companion Grandma nearly died at the hands of Lucas as well. Nadim had seen Grandma's baby approach the large male orangutan. Lucas suddenly lunged at the baby, but Grandma got in the way. Lucas grabbed Grandma and pushed her down into the creek water. The baby escaped. Witnessing this, Nadim began banging loudly with a stick and shouting. Lucas stopped and took his leave.

Lucas likely killed Jeffrey. Jeffrey clearly showed signs of becoming an adult male, but unlike other young male orangutans, he did not keep his distance from Lucas. Workers thought it strange when he no longer turned up for any of the feeding times. They eventually recovered his body off a trail near the feeding platform. The flesh was cleaned from his bones, likely eaten by a monitor lizard. Jeffrey's and Deh's skeletons are laid out under clear acrylic boxes, like coffins containing natural history museum exhibits. The Forestry Corporation issued their death certificates, indicating that both had died of natural causes.

Unlike in the past, when male orangutan violence against female orangutans was permissible as acts of nature and when violence between males required transfer, the official policy of the Forestry Corporation now is to refuse intervention. The newly appointed general manager, who had worked in Sarawak's national parks and biodiversity conservation for decades, has taken away that double standard. Current official policy at the Forestry Corporation disallows intervention "because it is natural" to kill and die and have sex regardless of the costs that living together might bear on those forced to reside there. The refusal to intervene in all cases of violence, save for the occasional salve surreptitiously rubbed on an open wound, characterizes this institution as a place of death, for both species and individuals.

Whose Lives Have Higher Priority?

The IUCN designated Bornean orangutans as critically endangered in 2016. Based on data from satellite imagery, conservation drones, and mathematical model projections, behavioral ecologists and conservationists have recalculated the number of orangutans at a lower figure for the most recent update of the Red List of Threatened Species (IUCN 2016; Wich and Kuehl 2016). A species' exact population, however, is less important than the fragmentation of its habitat: 4,000 orangutans in contiguous forest are better off than 4,000 orangutans in isolated pockets of forest.

Hundreds of other primates are on the list of critically endangered species. The populations of some number in mere tens, and for others no individual has been seen in decades. Of these critically endangered primates, which deserve to be on the Top 25 List of Most Endangered Primates?[2] Whose lives have a higher priority?

Layang had to confront this unenviable decision in the rainy season of late 2009 and early 2010. He and Nadim were among the Forestry Corporation employees tasked with rescuing endangered wildlife facing the imminent threat of drowning when Bakun Dam began operations. The dam had been mired in controversy since its inception, but an investment from peninsular Malaysia resuscitated the project. Nadim's team constructed and maintained a base camp, which they had to rebuild twice, once when the waterline rose and again when weather destroyed their camp. Layang's team trapped and released the wildlife they could find, with priority given to totally protected species.[3] The whole area smelled badly of stagnant water and the rotting leaves of drowned and dying trees.

A book jointly printed by the Forestry Corporation and the hydroelectric dam company commemorated the team's efforts (Pashal anak Dagang et al. 2013).[4] The published snapshots that stick with me include a sopping--wet nocturnal binturong in broad daylight, clutching for dear life to floating deadwood; a profile of a sambar deer, eye wide open as its head barely crests the water; the muntjac's and flying squirrel's wet bodies elongated in such confusing positions that it is hard to tell if they are alive or dead; and dripping-wet gibbons in traps, awaiting transfer to dry ground where they would have to re-establish new territories along with other gibbons who were also either displaced and relocated or who were already established there.

The 1,552 individuals rescued included fourteen totally protected species. Most of their fellow forest inhabitants were not saved. Nadim took pity on the ants, termites, and caterpillars crawling out of the trees as the trees became waterlogged and eventually submerged. With nowhere to go, the insects were slated to die there, just like the bees whose hives were about to drown. Rescue workers wore ski masks to prevent their stings. Macaques they found nearby risked stings rather than risk drowning. Totally protected hornbills' nests that were fused into tree cavities still had eggs in them. While adult birds and bees could fly away, their broods submerged in hives and nests were bound to die. Death stretched across the span of 42 km^2.

An event of this scale devastates far more than what ordinary people can observe. Three different men sleeping at different camps reported elderly women in white haunting their dreams. During the night, the men at different hours felt attacked while they slept and screamed out in fright, only to awake to their male crew members. The Forestry Corporation felt compelled to carry out two ceremonies to assuage ancestral spirits—a Christian ceremony as well as a "traditional" one. Perhaps the women were disturbed ghosts. Perhaps these very gibbons, binturongs, and hornbills were the nyarong of ancestors. I could not ask Layang how he felt about this, because the project was one of his last.

Layang reported for work at Lundu Wildlife Center the day after he returned from the Bakun Dam rescue. While performing husbandry that day, he felt ill, vomited, and, after cleaning the orangutan cages, went home sick. After three days of fever, Layang was admitted to the hospital. The medics suspected pneumonia, so they drew blood for testing. His fever persisted and he began slipping in and out of consciousness, eventually unable to recognize his loved ones. Layang passed away at the hospital.

Almost two hundred people, mostly former volunteers and including myself, joined an online group and contributed to a trust fund for Layang's family. ENGAGE set up the donation campaign through the website Virgin Money Giving—a not-for-profit arm of Richard Branson's Virgin empire. The site raised a little more than £6,000. The last donation was made eight months after Layang's death. His widow eventually used the funds to buy a *lading sawit*, a mono-crop field of oil palm. Her actions fulfilled the wry idea Layang once shared with me: "If I had the money, Juno, I would buy a palm oil plantation."

An orangutan at Batu Wildlife Center bears Layang's name. The orangutan's mother is the famously independent Wani, and he was the last of her children before she disappeared. By 2016, years had passed since Layang the orangutan was last seen. Perhaps he is bebas. Perhaps he is dead like his namesake.

Soon after Layang the man died in 2011, a story circulated in the alternative Malaysian newspaper, *Free Malaysia Today*: "Mysterious Deaths in Bakun Dam Area" (Tawie 2011). The article's use of quotes around the words "save" and "spirits" not so subtly undermined such terms:

> Ten people who were hired . . . to "save" animals from being drowned, following the impoundment of the Bakun Dam, are now dead. The dam was impounded in October last year. The 10 dead were allegedly killed by a mysterious disease. . . . He [the father of the last to pass away] believes they could have been exposed to infectious and deadly diseases such as melioidosis and leptospirosis (known locally as "*penyakit kencing tikus*" [illness from the urine of rats]). . . . Melioidosis is caused by a bacterium (*Burkholderia pseudomallei*) found in soil, rice fields and stagnant waters. Victims acquire the disease when the contaminated soil comes in contact with an abraded area of the skin. Leptospirosis is caused by bacteria called *spirochete* and is transmitted through contact with infected soil or water. . . . He [Bakun Community Safety Committee chairman] believes that the change of biodiversity could have resulted in the spread of the diseases. Local people believed that the "spirits" of the jungles are angry over the construction of the dam and the destruction of the surroundings.

What were the reasons for both scare quotes? Was the first pair of quotes calling attention to the incongruent scale between a rescued group of animals that could be handpicked by ten workers in relation to the devastation of Bakun Dam, which flooded a vast 42 square kilometers of forest? Or perhaps were the scare quotes meant to convey a question about the entire enterprise of rescuing animals? We may never know the full answer, but we are left with a feeling of skepticism, one similarly conveyed by "spirits." The quotes render them as alleged spirits, just as they denote an alleged rescue.

Two days later, the state-censored English newspaper *Borneo Post* offered another story, "No Mystery in Deaths." The first line of the article stated, "State Health director Dr. Zulkifli Jantan put an end to speculations on mysterious diseases and curses from angry spirits in the Bakun area by confirming that three deaths reported in the Bakun area recently were caused by melioidosis and leptospirosis in a press statement yesterday" (Sibon 2011). The cause of the workers' death was never in dispute: the first article mentioned both illnesses by name. However, the question of *why* they died was one that the Sarawakian state, through its newspaper, wanted to put aside. Like anthropologist Evans-Pritchard's (1937) well-known example of termites eating away at a granary post that causes its collapse, *Borneo Post*'s explanation of the mechanism of the workers' death—melioidosis contracted while working at Bakun Dam—does not explain the timing of their death.[5]

The answer to the question of why they died when they did might never be available to us, as never-appearing as the orangutans at Batu Wildlife Center, whose absence, with the lack of dead bodies as evidence to the contrary, indicates to the Forestry Corporation that the apes are wild. *Borneo Post*'s urban readership, already convinced that Bakun Dam should never have been built, can smugly feel confirmed in their opinion. Those urbanites who disdain competing cosmologies can smugly dismiss others' as superstition.

Still, for many, sudden deaths cry out for explanation, however elusive a definitive one might be. Witchcraft, curses, or illness all offer comforting reasons. The uncertainty of an answer does not preclude wonder.

I figure that Layang's death, with severe and rapid onset of symptoms, was probably caused by an acute case of melioidosis (Goris et al. 2013; Yingst 2014). For animals, including humans, melioidosis is extremely deadly since acute cases do not respond to antibiotics (Cruickshank 1949; Yingst 2014). Getting to the hospital earlier would not have changed anything for Layang and his family.

Microscopic Colonization

Had a form of colonization killed Layang? *Burkholderia pseudomallei*, the bacteria responsible for melioidosis, thrives in a living system at the expense of that living system's survival. *Burkholderia pseudomallei* flourish by destroying others' lives. Bacterial colonization works through "sacrifice," where some bacterial conspecifics give up their own lives so that others in their colony can flourish (Harshey and Partridge 2015; Hartigan n.d.: 14). Such colonization reinforces a hierarchy in which their own kind is more important than others.

The English-speaking world's understanding of melioidosis and the bacterium *Burkholderia pseudomallei* is due to imperialist geopolitics in Southeast Asia. The first report appeared in the 1910s with a case in Rangoon (Whitmore and Krishnaswami 1912). Initial samples came from the British colonies of Burma and Malaya, but were lost during the Japanese occupation (Cruickshank 1949; Green and Mankikar 1949). The bacterium, endemic to Southeast Asia and northern Australia, lives in soil. It can enter the body through contact with mud through one's mouth or a wound, or it can be transmitted airborne when soil is disrupted (Dance 1999).

During the Vietnam War, the U.S. Army and the Department of Defense initiated a study of melioidosis when a number of American soldiers were helicoptered into rural areas, fell ill, and died from it (Sheehy et al. 1967; Yingst 2014). Research carried out by the United States and colonial Britain entailed infecting guinea pigs, goats, and rhesus macaques with the bacterium (Caswell and Williams 2007; Cruickshank 1949; Yingst 2014). These animals either died rapidly upon infection, or were subsequently euthanized for postmortem dissection. In such dissections, researchers saw the growth of abscesses in the animals' lungs (Yingst 2014).

Vegetation is also susceptible to infection and subsequent death by melioidosis. Like all the guinea pigs injected with *Burkholderia pseudomallei* in Green and Mankikar's 1949 study, all the exposed tomato plants in Lee et al.'s 2010 study died within seven days. In a tomato plant, the bacterium infects the vascular system. Its virulence toward all kinds of complex eukaryotic living systems, including animals and plants, has led to its classification as a bioterrorist agent (Sprague and Neubauer 2004).

Such a classification suggests something inherently dangerous about melioidosis, but one can be a carrier without experiencing symptoms.[6] Indeed, before resuming my observations at Lundu Wildlife Center in 2009, I was required by Forestry Corporation to be tested for melioidosis in the event that I might be one such subclinical carrier. Of course I complied with my hosts' wish, though to my knowledge, no one else was asked to undergo such testing. I guessed I was singled out for testing because my mobility implied that I could more readily expose orangutans to illnesses I had picked up elsewhere than could workers whose mobility was limited to Sarawak. The presumption, it seemed, was that illness and pandemics came from elsewhere—whether avian flu, H1N1, or rabies.[7]

Like all forms of life, *Burkholderia pseudomallei* makes a place for itself within its biome, on specific surfaces of the Earth. Like Bornean orangutans, it is native to Sarawak. Recent whole-genome sequencing of *Burkholderia pseudomallei* samples taken from Sarawakian patients with melioidosis show unique mutations specific to Sarawak (Podin et al. 2014). Unlike samples taken from other areas of Southeast Asia, this particular strand has evolved to respond differently to gentamycin, an agent used in environmental tests for *Burkholderia pseudomallei*. Tests developed to detect the bacterium do not detect the Sarawakian strand. The bacterium is as unique to Sarawak as Bornean orangutans are to Borneo. While *Pongo pygmaeus pygmaeus* is endemic and endangered, *Burkholderia pseudomallei* is endemic and endangering. Along with multitudes of species, they make up the biome in Sarawak, where vitality for some can mean fatality for others. Sharing space with others capable of killing you is not a condition limited to Sarawak; it is a condition of living.

Acute melioidosis is usually detected through postmortem analysis, like myocarditis. Myocarditis can happen when an otherwise healthy young person gets a typical infection like strep throat or hand, foot, and mouth disease. *Streptococcus* and coxsackievirus proteins mimic human heart proteins. Someone dies of myocarditis when their immune system attacks their circulatory system's production of heart proteins necessary for their circulatory systems's operation. My familiarity with such a cause of death taught me this (Cooper 2009; Root-Bernstein et al 2009). Our human hearts produce protein chains like the bacteria that surround us, fill us, and make us who we are. We are porous selves within microbiomes and with microbes, sharing social lives even when we may not want to.

Knowing this, knowing the porous fragility of our bodies and our vulnerability to each other, how can we imagine a life with others, including endangered and endangering others? What could life be like for semi-wild orangutans if human caretakers no longer espouse isolated autonomy, but rather interconnected vulnerability? Could we embrace intimacy across difference, knowing that intimacy can be both harmful and enriching to us and to others?

We may well know that our bodies are always vulnerable to agents beyond our control, whether microbial and chemical, or intersubjective and global, but this knowledge doesn't matter much in the face of an untimely death like Layang's. His demise forces us to feel the pain of a lesson that no one wants to learn and yet that is the point of this book: Everybody we know will pass away. How well we die depends on others. Decolonizing extinction is one aspiration for what remains of our lives. This to me is about experimentally living together, feeling obliged to others, without a sense of safety or control that requires violent domination, and while being open to the uncertain possibility of experiencing harm from contact with others, even when that potential harm may be fatal. Do we have the courage to face up to this challenge?

Courage

Layang's job was humble in terms of salary and social status—the marks of success in twenty-first-century Sarawak. But he possessed legendary courage and confidence. There is a story about him hunting honey at Lundu Wildlife Center that I imagine he would have wanted others to know, so that we might remember his generosity and bravado.

At the longhouse one night, 'Kak and I ate dinner with her children at the kitchen table in her *bilik*. Over the meal of steamed red rice with dried tofu soup, sautéed pumpkin with greens, and, courtesy of her teenage daughter, sliced hot dogs stir-fried with sweet chili sauce, 'Kak explained that her husband was not present that evening because he had joined Layang, Apai Len, and Apai Julai on a mission to collect a honey hive in the forest of Lundu Wildlife Center.

Within those men's lifetimes, hunting for honey was not just a commercial practice. Rather, it was a *pengawa*, a term that in the Iban language connotes the significance of work and ritual (Sandin 1980). A quarter century before and on the other side of Borneo, Anna Tsing (2003) had accompanied a whole party of men, women, and children on such a hunt in the Meratus Mountains of South Kalimantan in present-day Indonesia. There, young men scaled honey trees and sang seductive lullabies to the bees as they traded golden fire for pillow-like honeycombs and woven mat-like brood combs (Tsing 2003). Benedict Sandin's (1980) ethnological studies dating from the 1950s, sixty years prior, described the practice of hunting in the darkness of the new moon; singing special verses to the bees; and distributing honey, combs, and wax among all the people who gathered at the tree as distinct parts of Iban adat, or customary law.

In 2009, no one else accompanied the group of men to the forest of Lundu Wildlife Center. Layang, a man in his late thirties, was the youngest in the party. No one openly tended the tree, which is one way to denote the right to scale the tree and take honey (Tsing 2003). Nor were the trees claimed with markings (Sandin 1980). Unlike in these ceremonial contexts, Layang did not have to climb a multitude of trees. The sole honey tree receiving the group's attention belonged to their employers, the Forestry Corporation. Technically, they were trespassing. This is likely why the party was so small and limited to subcontractors who worked at the site in the daytime.

While the men were on this mission, 'Kak told me a story about *indu manyi* (bees) and two brothers from the longhouse who died while trying to collect a hive. Still, she said she had no fear for her husband and his group. As she talked, her youngest child—the *bunsuh*—interrupted our conversation with tears and whines.[8] She wanted instant noodles instead of the dinner we were having. Sarawakian tastes, shaped by centuries of trade, didn't cut it for the bunsuh, whose palate seemed to have shifted toward the mass-produced foods manufactured for the global poor (Errington et al. 2013).[9] The next day Layang came by to give Ren, 'Kak, and their family their share of the honeycomb. 'Kak sautéed the bee larvae in a little bit of water, oil, salt, and MSG, and thanks to Layang, we all savored the taste of 'Kak's childhood. Even the bunsuh ate it happily.

I was not privy to the seductive magic Layang used to secure the honey and brood combs, the shares of which he distributed among the small group of participants.[10] I am not sure if Layang even knew beforehand what words would ease his way to the hive. What he did share with me was the bravery it took to carry out the act under tumultuous circumstances. On a walk weeks later, he proudly showed me the incredibly tall tree that he had scaled to obtain the hive. The alarmed bees began stinging him in a concerted attack as he accidentally dropped the hive to the ground. He descended as quickly as he could and jumped into the waterfall to escape. He laughed every time he retold the story.

Almost a month later, as we sat down on a bench for Layang's cigarette break, Apai Len walked by with a handful of cardboard egg crates. Whenever the mosquitos got bad during jungle skills training, Apai Len would burn the cardboard slowly so that the smoke would distract the mosquitos from their paths to our bloodstreams. But as Apai Len walked past now, he said the egg flats were for the bees. Layang giggled, prompting me to ask him if he gets scared (*takut*) when people say the word *manyi*. I must have offended him, because he snapped at me in English so that I would understand every word: "I'm not scared. I'd like to do it again. Bees tried to make a nest at my house. I just shoved them away. Don't think I have a phobia because of that accident. That's your experience." At first blush, this seems like straightforward pride in dominating others (Tuan 1984). Yet Layang's position at the top of such a hierarchy was tenuous, since the act of dominating bees exposed him to painful vulnerability. The attack could have been fatal, especially since his son is allergic to bee stings.[11]

It should not have surprised me that Layang's attitude about his conduct with bees paralleled the way he saw his conduct with orangutans. In neither case did Layang anticipate remaining safe. Closely encountering either species meant risking either stings or bites. Fear had no place in the relationship, and neither did safety.

Layang's life's work and Nadim's thoughtful actions toward Wani inspire the idea that decolonizing extinction means embracing risk and cultivating attentiveness. What does it take to be brave and accept the risk of infection or pain, not as a way to succumb to fatalism about the finitude of antibiotics in the face of virulence but as a way to live out an ethics of decolonization, one that knows all too well the fragility and diversity of all inhabitants of this planet? When we make conservation interventions, can we be less enamored with the proliferation of new life and be more concerned with the process of dying well? Can we come to terms with the future we have imposed on orangutans? Are we, like Nadim, able to shift our perspectives to imagine the thoughts and feelings of those who are different from us? Do we have the courage to face the uncertainty that arises in our relations with others, however fleeting our time together may be?[12]

When we begin such shifts of thought, we need to be open to possibilities we have not anticipated. The Iban-Chinese junior officer Cindy helps guide this. After a two-year absence, Cindy visited Lundu Wildlife. She was excited to see Lubok, whom she had lovingly cared for on overnight shifts after her 8 to 4 workdays, three days a week for an entire year. Upon seeing Cindy's joy and hope, Lubok did not have a visible reaction. The risk of loving and not having that love returned, the vulnerability of physically being subject to potential attack, and the uncertainty of what could happen when we relinquish control: that is the caring work of decolonizing extinction.

NOTES

INTRODUCTION: DECOLONIZING EXTINCTION

1 These figures are from Ibrahim et al. (2013).

2 For a discussion on the difficulty of categorizing native versus alien species, see Helmreich (2009). For the politics of identifying a species as alien or invasive, see Subramaniam (2014). The maps included in this ethnography are not meant as totalizing forms of material truth, but are offered as perspectives that can help readers navigate geographies that are possibly unfamiliar to them. For critical map studies in Southeast Asia, see the work of Thongchai Winichakul (1994) and Anna Tsing (1993).

3 The term *semi-wild* is mostly used by workers and rarely appears in the primatological literature, with the exception of the work of a Czech veterinarian working at the famous (or rather infamous) site, Bukit Lawang in Sumatra, where semi--wild orangutans socialize with both people and wild orangutans (Foitová et al. 2008). These orangutans are not "rehabilitant" because they do not fit the definition of rehabilitation offered by the IUCN (Beck et al. 2007).

4 A possible reason for separation was the fear of infanticide. This was not explicitly discussed, but implicitly conveyed when Layang and Lin fought about Ching's and Ti's captive conditions during pregnancy, which I discuss in chapter 5. Lin wanted the infants to be born in captivity and to separate the infants

from their mothers shortly after birth while Layang wanted to release them and have them give birth to their infants in the forest. Separating infants from their mothers was the process throughout the first decade of the 2000s. Their procedures changed with the general manager of the Forestry Corporation, who previously worked in conservation and who replaced a general manager whose career experience was in the private sector, working mostly in telecommunications.

5 By violence, I refer to the exertion of force or bodily harm. This sense of violence is more closely related to the work of Veena Das (2007), Margaret Lock, and Arthur Kleinman (Kleinman et al. 1997), despite their main interest being in specifically human suffering, than it is to Jacques Derrida (1997) and Elizabeth Grosz (1998), and the violence that occurs when things and objects referenced by words are forced into forms of representation. Nevertheless, these forms of force are connected through the sense of duress, strain, coercion, and pain conveyed by the word *violence*.

6 By "we," I mean a plural and inclusive we that includes you, dear reader, reading his academic monograph.

7 This is contrary to Yi-fu Tuan's (1984) point that human–animal relations are about domination and affection. His examples, such as dogs that have been bred for short noses to the point that they have difficulty breathing, do not lend themselves to considering animal agency or co-constitution. The animals in his argument are all victims and objects.

8 While vulnerability can be emotional or psychic, I am primarily interested in its physical manifestations.

9 In later chapters, I explain how "care" is not necessarily loving and pleasant, but rather something akin to the opposite.

10 To interface is to be co-present with others and to develop relations that form our subjectivities and make our worlds (Barad 2007; Butler 2004; Fuentes and Wolfe 2002; Massumi 2002; Middleton 2011; Riley 2007).

11 The population density is four orangutans per square kilometer, and that figure is similar to a few other sites in Borneo. Yet most sites with similar figures have been logged, thus likely contributing to the density of the population. For instance, only six of the eighty-one sites compiled in the first quantitative study that compared all wild orangutan field research sites in Borneo had equal or greater population densities (Husson et al. 2009). The average reported density in these studies is 1.81 per square kilometer. Logging activity plays a part in shaping the six sites with population densities greater than four orangutans per square kilometer. The authors find that "logging operations lead to inflated orangutan densities in neighboring, unlogged habitat" (Ancrenaz and Lackman--Ancrenaz 2004; Husson et al. 2009; Marshall et al. 2006; Morrogh-Bernard et al. 2003; Russon et al. 2001). Their point helps demonstrate that comparisons between "wild" orangutans in the field and rehabilitant orangutans in rehabilitation centers are not comparisons between pristine and wild nature on the one hand and unnatural, tainted culture on the other. Rather, what is observed in the "wild" is produced through logging and other human–animal–ecological mul-

tispecies encounters: felling large trees, lowering the availability of food sources, and the unintended growth of some plants over others are just a few outcomes of such possibilities. I further discuss the natureculture of the wildlife center in chapter 3. For natureculture, see Haraway (2008).

12 Before the bankruptcy of Global Limited, its CEO Ting Pek Khiing was the iconic figure of crony capitalism under the regime of Chief Minister Taib, who held power in Sarawak for thirty-one years, from the 1980s.

13 The irony was not lost on the park's workers who patrol the boundaries and have to gain permission to trespass Global Limited's property to be able to do their job of monitoring the western edge of the reserve boundary.

14 This is but one example of neoliberal governance (Sodikoff 2009; West 2006), yet one that more closely resembles new corporate forms vis-à-vis private–public partnerships like "government-owned NGOs" and volunteerism that provides services formerly provided by the state (Muehlebach 2012; Sharma 2006). For more on how emerging economies espouse liberalization see Cammack (2012).

15 The work of care speaks to a long tradition of feminist scholars who see that reproductive labor is work that is professional, even as it is casually compensated, and that it is gendered and racialized. See, for instance, the work of Anne Stoler and Karen Strassler (2002), Silvia Federici (2012), and Rhacel Parreñas (2001b). Additionally, feminist science studies scholars are now engaging a conversation about care in technoscience and how producing knowledge entails thinking with care. See, for instance, Aryn Martin and colleagues (2015), Maria Puig de la Bellacasa (2012), Astrid Schrader (2015), and Michelle Murphy (2015). In both strands, care was never happy, innocent, and devoid of power, inequalities, and forms of violence. Care always requires work.

16 Decolonization as a historical term usually references the formal exit of empire following World War II, either by diplomacy or violent liberation. I am compelled by Eve Tuck and Wayne Yang's (2012) point that decolonization is different from anticolonialism because anticolonialism is carried out by elites of the colonies who seek to replace old colonial masters with new postcolonial masters. Reynaldo Ileto (1992), in his classic essay "Religion and Anti-Colonial Movements," confirms their argument in that religion, and folk religion in particular, all across Southeast Asia—whether in Buddhist, Islamic, or Christian contexts—offered the means of anticolonial uprising. In the cases Ileto illustrates, religion offered an alternative hierarchy of power. Because these alternatives are posed as alternative hierarchies, it still imposes a hierarchy against which decolonial scholars write.

17 Decolonization is not like "decontamination" or "dehydration." There is no return to a pristine past or golden age. Rather, decolonization for many feminist scholars is about vigilance against domination, purity, and work toward new multiracial futures following contact. See, for instance, Ramirez (2008), Alexander and Mohanty (1997), Keating (2011), and Esquibel and Calvo (2013).

18 Many scholars of decolonization write about it in the present, following the collapse of the Soviet Union and with it the end of the Cold War. When the

resources of the Soviet Union became unavailable for Third World liberation, decolonization took on a meaning different from its reference to the formal end of empire and the emergence of political sovereignty. Most definitions of decolonization emphasize the liberation of colonized peoples, including people who have been officially liberated but lack access to their full rights as citizen-subjects. Such definitions explain that colonized peoples have been dehumanized and treated as animals. Implicitly, such a definition is attached to senses of human rights, justice, and liberalism. I find that these definitions of decolonization are made at the expense of recognizing social relations and obligations. The definition of decolonization that I am proposing, which is ultimately experimental in its embrace of uncertainty, is about liberation constrained by social relations and not about liberation constrained by justice on the basis of humanity.

19 A consideration of decolonization forces the question of *who* are the subjects. Actor network theory in science and technology studies has encouraged evaluation of how nonhuman actors shape the worlds in which we live (Callon 1986; Latour and Weibel 2005). As scholars in science and technology studies demand an expansion of politics that includes the nonhuman, like formaldehyde outgassed from the adhesives that bind the walls of temporary trailer homes or forests that decompose and grow anew in a poisoned ecology shaped by the American war on drugs fought in Colombia, nonhuman animals are often regarded as safely apolitical subjects in the everyday (Lyons 2016; Shapiro 2015). For instance, state-owned television in Malaysia regularly broadcasts *National Geographic* documentaries about wildlife, while videos of cats offer cute respite from disturbing news in troubled times (Chua 2017; Ngai 2012). Yet nonhuman animals are subject to politics. Donna Haraway's (1989) *Primate Visions* continues to be relevant in showing how such representations of nonhuman others are deeply entrenched in colonial politics and legacies.

20 Let us consider the concept of decolonization and what it would mean to decolonize institutions or conditions. For Ngũgĩ wa (1986), decolonizing the mind meant finding expression through African languages instead of mastering colonial tongues, to which postcolonial writers responded by demonstrating that colonial languages are not owned by colonial masters, but are instead subject to appropriation by the formerly colonized (Ashcroft et al. 1989). For Faye Harrison (1991), decolonizing anthropology meant countering the discipline's historical relationship to colonialism and imperial violence with a commitment by activist anthropologists to struggle against white supremacy, gender inequality, and economic injustice. For Linda Tuhiwai Smith (1999), decolonizing methodologies meant creating research projects by and for indigenous peoples to meet indigenous needs. And for Joel Wainwright (2008), decolonizing development meant adopting a postcolonial Marxist perspective and dissociating capitalism from development. All four see that decolonization entails incorporating peasant or indigenous perspectives that ultimately transform, and most importantly do not end, the institutions or concepts to be decolonized. These four scholars are more prescriptive than I wish to be. While Harrison and Wainwright both evoke the

language of experimentation, both are committed to a humanism informed by ongoing legacies of dehumanization. When decolonization is definitively limited to humans, it falls short of its potential to envision and enact expansive forms of justice. Hence, this is what makes the forms of decolonization evoked by feminist scholar Maria Lugones (2010) and education scholars Eve Tuck and Wayne Yang (2012) relevant to thinking about decolonizing nonhuman species extinction.

21 What Vicente Rafael calls "white love" is an exemplar of imperial domination that takes the form of benevolence, as is white imperialist feminism (Anderson 2006; Burton 1994; Rafael 2000). On the opposite end of that spectrum is the brutality of deeply unequal warfare (Fanon 1967).

22 Colonizing actions also sometimes take experimental form. Take, for instance, the experiment of forcibly settling Inuit people in stationary communities instead of allowing them to retain their nomadic lifestyles, which included even killing Inuit working dogs to make their people less mobile (Stevenson 2012). Tim Mitchell (2002) documents many more such experiments in colonial technocracy in Egypt, as does Tania Li (2007) in colonial and postcolonial Indonesia. The emphasis in my inquiry is not on experimentation in and of itself, but the experimentation of *kebebasan* and a sense of experimentation that is not predicated on unequal risk, in which harm is primarily experienced by research subjects and rarely by researchers. In this, I see a connection in the trope of the heroic scientist who risks "himself" for the pursuit of knowledge (Herzig 2005). However, a decolonizing sense of experimentation, I think, would be less self--congratulatory and would stem from a place of uncertainty instead of virtuous hubris afforded by privilege that is identified as rationality. This is evoked by Donna Haraway (2008) in the beginning of her chapter on shared suffering.

23 Some might consider that indigenous ways of knowing offer temporal frames that are incommensurable with such a sense of time as I am espousing. Following the work of linguistic anthropologist and insurance adjuster Benjamin Lee Whorf (1956) with Ernest Naquayouma, a Hopi man living in New York City, I'm inclined to disagree: time in Hopi is conveyed and grasped through adverbs and verbs instead of nouns in the standard average European languages of verb tenses. The time frames I describe here in English are certainly commensurable with Malay and Iban, which convey time not through tense but through adverbs.

24 The largest academic conference dedicated to feminist studies, the National Women's Studies Association, had the theme of "decoloniality" in 2016. This was met with criticism that decoloniality, like decolonization, is merely being used as a metaphor (Tuck and Yang 2012), that it means material changes to land ownership and citizenship. An example of the overextension of decolonization and decoloniality is the way Mel Chen's *Animacies* (2012) is regarded for its use of Chinook, an ergative language, to illustrate how hegemonic white American culture handles linguistic subjects. There are two major language systems in the world for expressing two arguments together, such as a subject and object in English. These two systems are ergative-absolutive and nominative-accusative.

Ergative-absolutive languages are found all over the world and are not geographically, culturally, or ethnically related, such as Chinook, Nepali, and Basque. Ergative languages treat grammatical subjects differently than nominative--accusative languages, which are also not geographically, ethnically, or culturally linked. Nominative-accusative languages are the most common in the world. Such languages include English, Korean, and the American indigenous languages of Miwok, Koasati, and Quechua. The argument that ergative grammar is relevant for people speaking the nominative language of English in English--speaking hegemonic American settler society is a stretch.

25 Mignolo (2000) interprets decoloniality as a "perspective of subalternity, from decolonization," yet that presumes static forms of power in which the subaltern is already known, delineated, and recognizable, which is contrary to Spivak's (1988) interpretation of subaltern studies. Another example of the overextension of decoloniality is the description of Frantz Fanon as "decolonial," which is an anachronistic application of the term. To argue that the identification of anachronism is an application of western chronotope fails to consider the ways time and chronology are conveyed by other means than the past tense. See footnote 23 in this chapter.

26 Lugones locates the violence of colonial domination in the creation of binaries between human/animal as well as men/women; however, readers of her work might get the sense that it romanticizes the precolonial as void of power and inequality (Lugones 2010). Decolonial scholars often misrepresent *postcolonialism* as a term that suggests colonialism is over (Arvin et al. 2013; Mignolo 2000). Postcolonial critics see ongoing colonial dynamics in all kinds of institutiuons, and they have fought ideas of imperial center and colonized periphery by emphasizing that all knowledge is local. See Anderson and Adams (2008) and their reading of Stuart Hall (1996) and Anna Tsing (1993). I use *postcolonial* in this text to indicate the time following official British colonization and the actions of the Malaysian state that inflict colonial relations. Following Yarimar Bonilla (2015: 4), the post of postcolonial is not just a periodizatgion, but a "mark of a transformed landscape of political possibility, distinct from a previous era of armed struggle and national revolution."

27 This is not to deny the violence of the year 1492 and its effects, which we continue to bear. Yet the problem lies in its resuscitation of a totalizing world systems theory that earlier anthropologists like Michel-Rolph Trouillot (1995) critiqued for its inability to recognize moments of uprising and resistance.

28 Likewise, Ursula Le Guin (2014) refuses to think of capitalism as an all-powerful hegemony. At the National Book Awards in 2014, she stated, "We live in capitalism, its power seems inescapable—but then, so did the divine right of kings. Any human power can be resisted and changed by human beings." Gibson-Graham (2014) and Roelvink and colleagues (2015) also refuse the hegemony of capitalism in order to show how other worlds are possible.

29 The Austronesian language family spans from Malagasy in Madagascar at the western edge of the Indian Ocean to 'Ōlelo Hawai'i in Hawaii and Rapa Nui on Easter Island in the eastern Pacific Ocean. It is thought that the rapid spread of

Austronesian languages was due to seafaring (Bellwood et al. 1995). What indigeneity means in Malaysia is contested, since the hegemonic nation-state centered in peninsular Malaysia claims indigeneity through the concept of *bumiputera* (sons of the soil) to the exclusion of Indian and Chinese diasporic Malaysians and at the expense of *orang asli* (who speak Austroasiatic languages, which are different from Austronesian languages) (Idrus 2010). As an Arabic speaker might guess, *asli* is a loan word that means original. In Sarawak, the racial demographics differ, where native Sarawakians are not *orang asli* but have been identified as *Dayak*, an umbrella term for indigenous people of Borneo who speak an array of Austronesian languages. Native Sarawakians demographically dominate the population, compared to 23 percent Malays and 27 percent Chinese, but native Sarawakians have less wealth (Jawan 1991; Leonie et al. 2015). Indigenous rights activists engage the term *orang asal* as an umbrella term for orang asli of peninsular Malaysia and native peoples of Borneo (Leonie et al. 2015). *Asal* is another loan word from Arabic, meaning origin or root. However, most Sarawakians who could fall under the term tend to self-identify with an ethnicity, such as Iban. All this information emphasizes that there is nothing inherent, intrinsic, or essential about indigeneity in this region of the world. For more on deep history, see James Scott (2017).

30 I do not espouse a rigid binary between two kinds of colonialism because both forms share modes of labor and resource extraction as well as monocropping. There are several examples from which to draw, including the United States and development of the Mississippi River for cotton and slavery (Johnson 2013), sugar plantations and slavery in Jamaica (Thomas 2011), and indentured servitude in tea plantations in Darjeeling, India (Besky 2014). However, a distinction is useful for my thinking since settler colonialism has a specific underlying malice of genocide based on racialization. While racialization happens in this context and other contexts of extractive and internal colonialism, oppression by way of dispossession is not clearly delineated across racialized distinctions.

31 Ibans are the largest ethnic group in Sarawak. They are indigenous and represent 28 percent of the population of the state.

32 Paleontologists agree that anatomically modern humans began appearing on Earth 200,000 years ago.

33 Pigs in particular are interesting to think with when it comes to the question of species standardization, being products of intensive breeding since the late nineteenth century. See the work of Gabriel Rosenberg (2016) and Alex Blanchette (2015).

34 Very few reported killing orangutans, and those who did, like one or two subjects, reported killing fifty to a hundred. As a political ecologist, I am suspicious of such a figure, and I wonder if the persons surveyed were boastfully exaggerating. Cultural anthropology and science studies have long investigated how quantitative data can obfuscate lived experience and generate problems specific to the data points such studies raise. Recent critiques include Sally Merry's (2016) work on sex trafficking, Diane Nelson's (2015) work on genocide, Vincanne

Adams's (2016) work on global health, and Michelle Murphy's (2017) work on population.

35 I'm indebted to Thom van Dooren's way with words for his phrase "edge of extinction." See van Dooren (2014).

36 Symbiosis, or the idea of biological organisms living together, is crucial for the contemporary theory of evolution that emphasizes coevolution instead of competition (Margulis 1981; Warinner and Lewis 2015; Woese et al. 1990). See Haraway (2016), Schrader (2017), and Tsing (2015) for feminist theorizations of microbes and symbiogenesis, or the theory of evolution in which all multicellular life derives from single cellular organisms. Enthnoprimatologists offer rich stories of symbiosis and naturecultures by researching primate interfaces (Fuentes and Wolfe 2002; Riley 2007). They situate their conversations in primatology, which until recently saw habituation, or the familiarity of a monkey or ape with a researcher, as a problem.

37 In her reading of Michel-Rolph Trouillot's (1995) *Silencing of the Past* and his challenge to the totalizing view of world systems theory through the village study, Vanessa Agard-Jones (2013) suggests that scholars look *from* the body, not *at* the body. From the body, can we look at the space between bodies? Indeed, I think a perspective from the body results in an ideal vantage point for perceiving what gets generated between bodies. For more on affect, see chapter 2 of this volume and Rutherford (2016).

38 I was not local, but my hair and skin color meant that Ching had the power to reduce me to my appearance. My diasporic personal history as a Filipina American had no bearing on her. I was vulnerable to her attack; so were the female staff members and any of the ecovolunteers working at the site with similar features. I made efforts to never look at Ching directly, to turn my body away, even when taking focused notes about her. I wanted my body to convey avoidance, even in the tight spaces in which we were confined. In so doing, I figured that I could at least behave in ways that might have been more acceptable to Ching (Knott et al. 2008).

39 For more on how race is a social formation that is culturally specific and mediated through the interpretation of bodies and associated signs, see Omi and Winant ([1986] 2015), John Hartigan et al. (2013), and Amade M'charek (2014).

40 'Kak had very little risk of running into Ching because 'Kak's activities were concentrated near the quarantine, which is a site that the orangutans associate with tranquilizing blowguns.

41 Few primatologists use the term *semi-wild* in their publications concerning orangutans. One example is the veterinarian and PhD Ivona Foitová, who used the term to reference the population of semi-wild orangutans at Bohorok in Sumatra, Indonesia. Bohorok was a rehabilitation center founded in 1973. Rehabilitation efforts were later abandoned because they risked the health of wild orangutans in the area. The multigenerational population that resides there are aptly described as semi-wild, especially since their behaviors are markedly different from wild and captive orangutans.

42 This interface of extinction is similar to the one experienced by conservation

biologists interviewed by Irus Braverman (2015) and shadowed by Thom van Dooren (2014). In both cases, they involve animals with conservationists seeking to save their species. These are scientists with authority to determine how to encourage life and at what expense.

43 Foucault's ideas of biopolitics have been indispensable for understanding the management of life in animal studies and queer studies. The work that occurs in Sarawak's wildlife centers shows the limits of biopolitics. While we see how captive breeding is a biopolitical act in that it is about the attempted management of a population, the very agency of orangutans questions the limits of what we can consider to be "management." The empirical evidence I present in these pages show that biopolitics is not enough to consider how orangutan rehabilitation is a life and death matter. We cannot continue to privilege the relations between discursive power and material bodies. Rather, we need to feel a much bigger world, one deeply uncertain and less contingent on any kind of distinction between mind and body.

44 People abducted from Ambon, Borneo, and other islands were sold into slavery in Batavia on Java. Their descendants are known as Cape Malays in South Africa (Allen 2014; Vink 2003).

45 I offer what I think the referent of the words might have been, although I must admit that in doing so, I risk a form of what Projit Mukharji (2014) calls "retro--botany," or in this case "retro-zoology," where I could be erasing the ontological concepts inherent in what each of those actual terms conveyed.

46. The description Bontius offered said that female orangutans were modest in that the one he saw covered her genitals. He explains that their origins were due to native women's bestial sexuality. He writes, "The name given to them is Ourang Outang, which means man of the woods, and it is said that they are the result of the lust of the women of the Indies, who slake their detestable desires with apes and monkeys" ([1642] 1931: 285). A version of this ideology appears around a century later in Thomas Jefferson's writings about "oranootans," which became a generic word for apes, and for which he proposed that they prefer black women over their own species. See the work of Brigitte Fielder (2013, 2017).

47 Please see chapter 5 for further explanation about the term *autonomy* and its utility. The figure of 11 million years is based on the hypothesized timelines given by Wood and Harrison (2011). For thinking critically about the Family of Man, see Haraway (1989).

48 Frantz Fanon, for instance, exemplifies this with the life history that informs his work, from living in France as a black medical doctor from the French Antilles, to his work as a militant revolutionary in Algeria violently opposing French imperialism (Fanon 1965, 1967, 2008). Benedict Anderson (1998) makes the point that nationalism is made possible by the specter of comparisons, a line from the Filipino nationalist Jose Rizal's novel *Noli Me Tangere*, in which the protagonist, a young mestizo, comes to see his homeland in a different way and is able to project a different future following liberation from colonialism. I think a similar specter of comparison is at stake, but not one that venerates Europe and nation-

alism as the zeitgeist, but rather one that compares other colonies struggling for liberation to each other.

49 The oldest human remains found in Borneo are dated to about 37,000 years ago, from the Late Pleistocene.

50 People of Borneo's interior had extensive trade networks, as evidenced by inherited ownership of Chinese pottery from the 1400s (Padoch and Peluso 1996). Sarawak Museum holdings include locally excavated Chinese ceramic sherds, mostly from 1000 CE to 1200 CE, but even from 800 CE, which evidences an older coastal trade. Precolonial Sarawak was not necessarily idyllic. Various groups waged war regularly, especially in the 1600s, according to the work of Benedict Sandin (1967, 1980).

51 This was primarily done by means of headhunting, which was a common practice in Southeast Asia (George 1996; Rosaldo 1980).

52 This was about forty years after Penang became a British free-trade port in the Strait of Malacca.

53 The lack of a historical or ethnographic record that corroborates a possible story of deification is different from the case in Hawaiian studies. Anthropologists Marshall Sahlins and Gananath Obeyesekere famously debated the possibility of ontological and epistemological commensurability and incommensurability. Anthropologist of the Pacific Sahlins (1995) held the position that Captain Cook was venerated as the god Lome by Hawaiians, while Obeyesekere (1992), a psychological anthropologist influenced by Freudian psychoanalysis, held the position that Captain Cook's veneration was a colonial fantasy. While the idea that Captain Cook was a god is likely unappealing to contemporary antiracist and anticolonial scholars, agreeing with Obeyesekere would mean claiming a human universal rationality and Western cosmology at the expense of considering alterity and incommensurable cosmologies.

54 They also fashioned themselves in the model of the sultans from whom they usurped power: they wore yellow, which regionally was limited to sultans.

55 In his proclamation known as the Nine Cardinal Principles, the eighth principle read as follows: "That the goal of self-government shall always be kept in mind, that the people of Sarawak shall be entrusted in due course with the governance of themselves" (Yong 1998: 153).

56 In 1961, after fifteen years of direct British rule following World War II, the prime minister of the newly decolonized Malaya across the South China Sea expressed the idea of forming the federated nation-state of Malaysia to the prime minister of Britain as a way to counter the rise of communism in Asia. Britain had just spent more than a decade waging anticommunist war in Malaya, which entailed such tactics as indefinite detention, forced migration, and the first weaponized use of Agent Orange (Khalili 2013; Komer 1972). Sarawak would be the new nation-state's largest territory. The British Crown set up the Cobbold Commission to assess Sarawakian support of this plan (Chin and Langub 2007). Although they found that opinions were split three ways between yes, no, and maybe, the Crown interpreted this as a unanimous yes. This led to war be-

tween Malaysia and Indonesia and a communist insurgency that was ultimately suppressed following the genocide of Communist Party members in Indonesia (Yong 1998, 2013). Sarawak was a politically independent and sovereign nation for less than two months: in 1963, between July 22, the day Sarawak was declared self-governing by the Crown, until September 16, the day the Malaysia Agreement was signed by Sarawak's first Chief Minister, which officially formed the nation-state that we now know (Leigh 1974).

57 At the fiftieth anniversary, Sarawak began celebrating its own day of self-governance, complete with dramatized reenactments involving white actors playing British colonial officials. See Harding and Chin (2014). Around that time, in the 2010s, bumper stickers and T-shirts with the slogan "Sarawak for Sarawakians since 1841" began appearing throughout the city of Kuching. Ironically, assertions of Sarawakian state autonomy in relation to the Malaysian federal state were articulated through Sarawak's despotic days ruled by a white autocrat. My friend Kelvin Egay, an anthropologist who is Kelabit, once joked, "Why not 1840, when we were all killing each other?" His joke was a funny reminder to refrain from romanticizing the past.

58 At the time of writing, the federal state has yet to adequately address the scandal in which Prime Minister Najib tun Razak, head of the ruling party, appeared to have siphoned US$700 million for the 1Malaysia program to his own personal bank accounts (Domínguez 2015). The U.S. Department of Justice is seeking to recover one billion dollars out of $3.5 billion dollars allegedly misappropriated from the 1Malaysia fund (Department of Justice 2016; Federal Bureau of Investigation 2016). These funds were meant to provide the poorest Malaysians with services, including health care. Despite the end of the Internal Security Act, which enabled indefinite detention for political opponents, continued political repression means that Sarawakians and other Malaysians can face a two-year prison sentence for such minor offenses as displaying a sticker that reads #tangkapnajib, a social media call to arrest the current prime minister ("Woman Queried" 2015).

59 Such a hierarchy need not be rooted in colonialism. What comes to mind is Jean Langford's (2013) description of the Khmer Rouge's cruelty in Cambodia, which Cambodian refugees in the United States described as being treated like animals.

60 The sense of rehabilitation evoked by orangutan rehabilitation is more closely aligned with ideas of prison rehabilitation and its relationship to recidivism than with medical rehabilitation following injury or disability (Wool 2015). In some respects, the rehabilitation center for orangutans stands in direct opposition to João Biehl's (2005) idea of zones of abandonment, because these are not places where the animals are left to die, but are subject to direct and indirect interventions on how they are forced to live.

61 The concept of *agencement* is extremely useful here. Agencement has previously been translated as assemblage, which has been the shorthand for talking about the convergence of multiple agents enacting changes in the world. However, the

word *assemblage* conveys a strongly mechanistic property. Agencement, instead, conveys the circumstantial convergence of such momentary groupings. See Phillips (2006) and Roelvink et al. (2015).

62 Such enclosures offer comparison to other "spaces of containment" like reservation land territorialized by the American settler colonial state (Goeman and Denetdale 2009).

63 *Tʃari makan* (tcha-ree ma-kahn).

64 Jason De Leon (2015) profoundly engages the material objects left behind by migrants in the Sonora Desert, which has been weaponized by the U.S. Border Patrol's Prevention through Deterrence.

CHAPTER 1: FROM APE MOTHERHOOD TO TOUGH LOVE

1 Tom was the British project manager of the commercial volunteer company supporting rehabilitation efforts at Lundu Wildlife Center.

2 Throughout this chapter, I identify the workforce employed by the museum by their first names. Their last names are the first names of their fathers. The respectful form of address is their first name alone.

3 Alfred Russel Wallace raised an infant and was the first person to write about it in his travelogue, *The Malay Archipelago*. However, he raised the infant after having shot and killed the mother. See Wallace ([1869] 1986). At the time, Wallace was shooting orangutans for scientific collection, which for him was a source of income. Extinction was not perceived as an issue until the twentieth--century conservation movement, in which Barbara and Tom Harrisson were key figures.

4 Surnames among Malays and Dayaks are patronymic, based on the first name of the father. Malay names follow the Arabic patronymic pattern where *bin* or *binti* designates son or daughter of a father. Dayak names are designated with *anak*, or child, of a father. Barbara Harrisson published Bidai's full name, with his father's title, in her journal article in *Oryx* about their efforts in Bako National Park (Harrisson 1963).

5 This ambiguity led to the accusation of theft at the end of Tom Harrisson's term, which coincided with official decolonization. Tom Harrisson was officially banned from Sarawak following decolonization. See Heimann (1998).

6 According to Heimann (1998), the Harrissons were forced to leave because of politics: Tom Harrisson offended too many of the new officers filling colonial posts previously held by European expatriates. This culminated in an accusation that he stole Sarawakian artifacts. Heimann suggests that his accounting records were chaotic, and purchases he made as curator of the museum were ambiguous as to whether they belonged to the state or to him. On account of his reputation, Barbara Harrisson was not allowed to enter Sarawak in 1967 when returning from Sabah, where she had left the last two orangutans that she had personally raised, only to have one of those orangutans be shot and killed (Harrisson 1987). This was an extremely difficult confrontation. In my conversation with Barbara Harrisson decades later, she became visibly upset and began raising her voice as

she explained that the man who shot that juvenile orangutan was not punished by the person in charge of the wildlife center, G. S. De Silva, because he would not punish a fellow Ceylonese on the account of a British colonial in the very first year of decolonization. Speaking with De Silva about this incident years later, in a shopping mall in an Asian-majority suburb of Toronto, he spoke with immense regret that an orangutan, one personally hand-raised by Barbara Harrisson, was killed so quickly in an accidental shooting. He explained that at the time, he only had two employees, neither of whom could care less about the animals in their charge.

7 When I spoke with her in August 2006, she spoke of herself as a "British colonial." Stated with a persistent German accent, the tone of her voice did not betray irony.

8 See file MU/444/1, dated January 19, 1967. The Wenner-Gren Foundation has funded anthropological scholarship since the end of World War II and has thus shaped the discipline (Lindee and Radin 2016).

9 Barbara Harrisson resonates with me as a queer figure because of her actions against the heteronormative family and her expression of a queer ecology of motherhood (Mortimer-Sandilands and Erickson 2010; Povinelli 2001). However, I would not want to anachronistically describe Harrisson's behaviors or sentiments by calling them queer. Historically speaking, her prioritization of museum over household would likely be regarded as simply eccentric.

10 Barbara Harrisson quickly dispels any question about her possible longing for children by employing humor. She writes, "As the years went by, she [Dayang] stayed on with us in spite of this defect, hoping against hope. Similarly, grateful to her faithful service over the years, I refused to see that she could not see well with her one eye, hoping against hope. For the house was often too dirty: cobwebs and dust accumulated freely, which she would take away with a broad smile when I pointed them out. An ideal relationship!" (1987: 37).

11 The American equivalent of *Kinderpflege-Lehrbuch* was Mary Mills West's *Infant Care* (1914), which she wrote for the Children's Bureau and the U.S. Department of Labor.

12 Structural exclusion has made the home a gendered site of knowledge production and innovation for women excluded from formal institutions (von Oertzen et al. 2013).

13 This is another example of a nonhuman form of Foucauldian biopolitics.

14 Eve resists, which is more than what could be said of the dominated species described by Yi-fu Tuan (1984). His examples include goldfish manipulated through breeding so that their eyes are too big for their sockets and the pug breed of dogs whose shortened faces, which were the product of human artificial selection, hamper their ability to breathe. In Tuan's vision of domination, animals are objects, not agents; they are acted upon and not actants themselves. Likewise, Harry Harlow's infamous experiments in which he separated newborn rhesus macaques from their mothers to prove the need for primal attachment called forth a short-lived version of a similar kind of domination—short in the sense

that it occurs within individual lifetimes and not over multiple generations; calling it short does not discount the long-term effects likely felt by the individual over time. Animals have the capacity to feel traumatized; see Braitman (2014) and Langford (2017). Unlike Tuan's vision of animal objectification, Eve is an actor and subject whose initial refusals should be read as acts of resistance.

15 Rice is a cultivar that requires cooking and would never be in a wild orangutan's diet. Harrisson in her book does not explain why she added rice. One could guess that it was to thicken milk that was likely powdered, or perhaps to add calories to their diet. In person, she explained that the lack of resources meant finding different nutritional sources from what could be duplicated in Sarawak's forests.

16 Harrisson perhaps had a female-bodied predecessor in the form of Delia Akeley, whose memoir from 1930, described by Donna Haraway (1989), detailed how Akeley's husband, the taxidermist and naturalist Carl Akeley, depended on her during a field expedition in what was then the Belgian Congo to nurse him to health and to seek out specimens. However, there were a couple of pressing differences between Barbara Harrisson and Delia Akeley: Akeley accompanied her husband to the field. Harrisson ventured on her own with men from the Sarawak Museum, whose responsibility was to assist her in her research. Akeley sought to kill and collect. Harrisson sought to observe. While Akeley's expedition could be understood as analogous to collection and mastery in a colonial era, Harrisson's mission was about finding inspirational models for survival following contact. My choice of words intentionally connects to Mary Louise Pratt's (1992) idea of the contact zone as the space of colonial, transcultural encounter, which Haraway (2008) uses in her understanding of the contact zone between companion species. Harrisson preceded Jane Goodall's research, which began upon her arrival in Tanzania in 1960. It was seven years before the vision of a young woman studying apes captivated the National Geographic television audience in 1965 with the broadcast of a special about Jane Goodall. Harrisson also finds a predecessor in the life work of Mary Cynthia Dickerson, editor of the American Museum of Natural History's journal between 1910 and 1920 and curator of herpetology as well as of woods and forestry before she was committed to an insane asylum where she spent her final years (Fabian 2013). Historian Ann Fabian has been tracing Dickerson's fascinating life and work at the tenuous borders of knowledge-making in the early twentieth century.

17 In the same way that my work as an intern zookeeper entailed picking up droppings at Oakland Zoo and placing them inside the freezer, which thereby transformed muck into a stool sample, Bidai's skinning of creatures for the museum transformed corpses into "specimens." See Parreñas (2016).

18 With Ina, Bidai was not as isolated and vulnerable to exploitation as domestic workers often were or are.

19 For more on animality, see Chen (2012), Kosek (2006), and Rothfels (2002).

20 Ann Stoler (1995, 2002) can help make sense of the anxieties around racial mixing and the policing of lines between civilized and savage in the rearing of European children in the colonies. Harrisson's musings invite us to think of her

care for orphaned infant orangutans through Stoler's critical postcolonial lens. The zoo for Harrisson is an institution of constraint and subject formation analogous to her own understanding of English boarding schools as sites of unloving constraint and subject formation for young colonial elites. Her opinion of such schools was shaped by her husband's experience as a boarding student in elite institutions like Eton. Harrisson sees that both the boarding school and the zoo are modern necessities—colonial parents "have to," or in other words *must*, send their children to boarding schools—if the children are to have viable futures in colonial society.

21 The animal handler Layang, upon seeing figure 1.1, pointed to the vulnerability of the infant. He felt that keeping it warm, especially during the wet season, was necessary yet difficult without the infant's mother. He said that the orangutans *always* die after we separate them from their mothers when they're still young. This particular infant did not have a name nor a year attached to it. Thus it is difficult to ascertain whether or not the infant survived. In Harrisson's monograph, all of the infant orangutans survive, despite the threat of human–orangutan illness transfer, worms, and malnutrition prior to their arrival. Early in her monograph, she states, "Experience has shown that Orang babies kept in long-houses or private households as pets only had a slight chance of survival. They died in a matter of weeks if not protected against *human* diseases and not given the right kind of food" (1987: 45). Those orangutans that do not survive are not named in the book nor in the archival records.

22 Harrisson's sense of responsibility offers a point of comparison to Naisargi Dave's (2014) analysis that eventful witnessing leads to a sense of responsibility toward animal others. Unlike Dave's subjects, Harrisson has a more complex relationship since "care" is not dressing wounds or ending pain, but tied to painful resuscitation via force-feeding and colonial tutelage toward eventual independence.

23 The project is described in the preface to the 1987 printing of Harrisson's *Orang-Utan*. The space allotted to them to continue their experiment was inadequate, and it did not have enough food to sustain their independence. The space was open to the visiting public, and an incident had occurred in which people were injured after running away from the growing juvenile orangutan Arthur. Arthur and Cynthia were thereafter transferred to the newly opened Sepilok Orangutan Rehabilitation Center. There, Arthur was shot and killed by one of the two workers at Sepilok. See chapter 6.

24 See Daniel Rosenberg and Susan Harding (2005).

25 See the introduction and chapter 5 for more on autonomy. In brief, *autonomy* here references biological autonomy and not the atomistic autonomy of liberalism.

26 Teasing about Apai Len's effete ways, which he accepted with shared laughter, evokes Tom Boellstorff's (2004) distinction between heterosexism and homophobia, in which people might tolerate sexual and gender diversity, but they espouse a heterosexism in which they socially police gender performance. See also Jia Hui Lee's unpublished paper that received AQA's 2011 Kenneth W. Payne Student Prize.

27 Working with rehabilitant orangutans meant being tough to volunteers as well. Even though all the volunteers signed an extensive list of forbidden acts that would get them kicked out of the program, a volunteer violated the agreement—she was touching the orangutans. When Ben saw her, he was shocked—too shocked to speak. He asked her friend, "Could you please tell her to not touch the animals?" He told Tom later and Tom said, "You have to tell her to not do it. You have to be tough." As he retold this story to me, Ben said, "But I can't, I'm kind!" For Ben and other keepers, being kind was ostensibly different from being "soft."

28 Race and gender remain material-semiotic signs, even as technoscientific innovations like genetic research are often deployed to pin them down to matter. Race, like bodies and like gender, is multiple (Haraway 1989; M'charek 2013; Mol 2002).

CHAPTER 2: ON THE SURFACE OF SKIN AND EARTH

1 To interface is to be co-present with others. Interface, for instance, recalls the way primatologists Erin Riley (2007) and Agustin Fuentes and Erin Wolfe (2002) discuss the shared interface between humans and other primates, which is their term for talking about sympatry or shared geography. Others might think of Emmanuel Levinas and the face-to-face relation of ethics, which in Levinas's world is limited by "human exceptionalism," since he felt that only human faces deserve ethical consideration (Butler 2004). Deborah Bird Rose (2004) reads against the grain but nevertheless recognizes that we face ethical relations with nonhuman animals.

2 See the Hornbill conference proceedings from December 1999 (Forest Department 1999).

3 Previously, I have considered the production of affect between orangutans, humans, and other bodies (Parreñas 2012). However, I am compelled by the feminist Gens Project to consider that words matter, especially words that work, namely the verbs *to produce* and *to generate*. I am inclined to think of affective states as generated, not necessarily produced, following Bear et al. (2015), because gens, which gives us the basis for the word *generate* and *gender*, gets us to consider the "appropriation of human and nonhuman life forces by social forms." It makes me realize that production connotes far more intentionality and mechanistic modalities than I wish to connote.

4 For more on the contact zone, see Pratt (1992) and Haraway (2008). Demonstrating the philosopher of science and psychologist Vinciane Despret's (2004) concept of "anthropo-zoo-genesis," we can see that in the physical space of the encounter, Ching the orangutan "could make human bodies be moved and be affected." Despret uses examples from famous science experiments, such as Clever Hans the equine mathematician and Rosenthal's naïve student researchers studying "dull" rats, to show that what is pejoratively disregarded as "bias" is the result of a mutually transformative and embodied relationship in which animal research subject and human researcher become attuned to one another. Such

affective encounters of anthropo-zoogenesis inspire Donna Haraway (2008) when she pulls from Karen Barad's (2007) idea of "intra-action" and from Isabel Stenger's reading of Alfred North Whitehead's idea of "mode of coherence" in her interpretation of postcolonial theorist Mary Louise Pratt's "contact zone." In the contact zone that Jake Metcalfe cleverly describes as "Harawayenne," following the collaboration between Haraway and her dog Cayenne, the act of connecting produces subjects that are in relation to each other, so that "each partner [becomes] more than one but less than two" (Haraway 2008: 244). The partnership possible across species, whether short-lived or durable, is not the romantic and Aristotelian cliché of becoming complete as a pair. Rather, in this world-making project, selves are never complete wholes; we can never be isolated "ones" that add up to a whole number when we congregate. Instead, we subjects are always fragmented and in a state of becoming. This sense of ontology finds its roots in feminist science studies and the privileging of "body" as a means of partial knowledge. The idea of bodies in feminist scholarship is broadly construed to include people, places, and other assemblages.

5 Volunteers were forbidden from taking selfies at the site. This reinforced how much embodied presence mattered for the act of volunteering, that it could not be represented and shared with friends who might want to vicariously experience the embodiment of visiting such a site.

6 Historian and queer theorist Gabriel Rosenberg writes about the gendered aspects of husbandry in his forthcoming book, *Bad Husband*.

7 Two of the new orangutans were infants born on site, and one was James, an adult male transferred from Batu Wildlife Center. I introduce him later. Three orangutans had been there the longest: Efran, Ching, and Ti were part of the first population transferred from Batu Wildlife Center to Lundu Wildlife Center. Lisbet was born at the site in 2002.

8 Conservationists' estimates of the wild orangutan population in Sarawak range from 1,500 to 2,000 (Chan 2009). The number of rehabilitant orangutans is low: thirteen at Lundu Wildlife Center and twenty-six at Batu Wildlife Center as of 2010. However, rehabilitant orangutans are the primary source of public media on orangutans, and visits to rehabilitation centers offer a chance to see orangutans in their natural setting. Orangutans in the wild are very difficult to find since they are arboreal and are the most solitary of the great apes. Batu Wildlife Center gets thousands of day visitors every year while Lundu Wildlife Center has many more volunteers than visitors. Batu Wildlife Center's setting appears more "natural" than Lundu Wildlife Center to visitors.

9 The concept of environmental stewardship is especially racialized in settler colonial societies. See David Hughes (2006b). It is also gendered through the idea of husbanding resources.

10 This particular conversation happened in English. Layang and I would speak either Malay or English.

11 Zookeepers from Australia designed the orangutan area of Lundu Wildlife Center, and its structure resembles other "world-class" zoos. Three outdoor enclo-

sures are connected to one building where the orangutans are kept at night. The building, known as the night den, houses eight cages. In the interior of the building, a corridor connects the cages. The cages have a species-size gate that opens to the enclosure and is operated from the corridor. The cages also have a gate door that opens to the corridor so that people can enter to clean the cage when the animal is in its enclosure. The corridor also has doors that open to the three enclosures. In Gas's night cage, the mechanism that would open the orangutan-size gate to her enclosure from the corridor was broken, so she had to be carried or led through the corridor and brought to her enclosure through the door.

12 Tim Ingold's anthropology of lines argues that European colonialism is not the imposition of lines or linearity among nonlinear thinking and people. Rather, it is the attempt to impose one kind of line upon other kinds of lines. Lines in this sense are not the mathematically and logically derived distance between two points, but are paths (Ingold 2007).

13 We are caught in what Heather Paxson (2013) calls "microbiopolitics," in which our human connections with bacteria are subject to the greater question of how we want to live.

14 Mut's vocalization resembled what MacKinnon (1974: 63) describes as "crying and screaming" and "fear-screaming."

15 Orangutans are not physiologically adapted to walking (Winkler 1995). Yet rehabilitant orangutans often walk upright. Examples of rehabilitant orangutans walking on the ground can be seen in the photographs of the book *Orangutans: Wizards of the Rain Forests* (Russon 1999).

16 Linguistic anthropologists following Jakobson (Caton 1987) point out that communication is more than just a tool for logical, referential meanings. What would happen if animal behaviorists and evolutionary psychologists thought like linguistic anthropologists when studying human–animal communication? The difficulty of ascertaining meaning and concretely defining nonhuman primate gestures became widely known in October 2010, when the journal *Cognition* retracted Marc Hauser et al.'s (2002) article, "Rule Learning by Cotton Top Tamarins," because its "research data do not support the reported findings." Instead of locating logics in human–animal communication, what happens when we become interested in finding feelings that evoke responses? Recent research in animal cognition investigates this in exciting inquiries regarding attentiveness (Liebal et al. 2007; Povinelli et al. 2003; Tempelmann et al. 2011).

17 I heard this statement when I accompanied a group of volunteers on their first day at the center as they toured the facilities. Being "really here," facing an endangered orangutan in Borneo, shows the way in which animals endemic to Borneo stood to represent Borneo as a land of jungles, wildlife, and nature for some volunteers. It stood in contrast to the "there" of urban, postindustrial Britain, regardless of that particular ape's captive condition. The irony that volunteers could hop in a van and get toast and marmite within an hour was perhaps lost on them. The point for them was that they were in a tropical forest, encountering members of an endangered species that they would otherwise never encounter.

18 For more on crittercams, see Haraway (2008). Rhacel Parreñas addresses how the distance between mothers working overseas as domestics and their children left behind in the Philippines creates feelings of jealousy and longing through an unintentional denial of intimacy. Despite mothers' efforts to preserve relations through frequent communication via text message and mobile phone, their children often nevertheless express resentment (Parreñas 2001a, 2001b). Technology mediates intimacy, like heartbreak on social media (Gershon 2010) or customers sexually harassing call center workers (Mankekar and Gupta 2016). Yet these technological mediations deny the experience of physical co-presence, which is one of the prized values in commercial volunteerism. As the bodily intimacy of nursing in an oncology ward in Botswana has the power to rehumanize patients dehumanized by cancer, the bodily intimacy of sharing physical presence with orangutans and other Southeast Asian animals emphasizes intersubjective relations (Livingston 2012; Parreñas 2016).

19 Orangutans are the only arboreal great ape and are not physiologically suited to knuckle-walking like gorillas, which are terrestrial. However, male and female rehabilitant orangutans mimicked the humans around them and would "walk" upright on their hind limbs, folding in the digits of their hind limb "hands" so that their palms could be flush against the floor or they would curve their hands in, palms facing each other as they walked. The forelimbs would drag in this process.

20 I had been in the night den with them taking notes concerning the morning animal husbandry routine up until that moment. I had stepped out for what was to be a brief moment when Efran was outside in his enclosure. Thus I had not directly witnessed the commotion, although I was just outside the night den. This incident made me keenly aware of how mundane co-presence can give way to spontaneous intensity.

21 Galdikas writes: "I began to realize that Gundul did not intend to harm the cook, but had something else in mind. The cook stopped struggling. 'It's all right,' she murmured. She lay back in my arms, with Gundul on top of her. Gundul was very calm and deliberate. He raped the cook. As he moved rhythmically back and forth, his eyes rolled upward to the heavens" (1995: 294).

22 Although the enclosure walls of Lundu Wildlife Center were actually built by ethnic Iban women laborers, the volunteers were unaware of how local women's labor helped build the place. All the workers with whom the volunteers worked were men. Since the years when the center was built, manual labor has become gendered as masculine, as it is in the home countries of the volunteers. This is not to say that labor that is gendered feminine does not entail physically demanding work. In Sarawak, working outside has come to be seen as unfavorable, especially by women.

23 Muriel entertained this question in a formal interview that I had with her in 2010.

24 Secular volunteer tourism offered a polar opposite to the cleanliness of "ethical capitalism" and global charity, yet they both operate on the same ethos of "im-

provement" through capitalist interventions (Kapoor 2012; Li 2007; Žižek 2010). The work of secular commercial volunteers contrasts with the sense of indebtedness Japanese aid workers in a Shinto-based NGO cultivated among themselves in Burma (Watanabe 2015).

25 Muriel's work in Scotland appears to convey what Hardt describes as affective labor (Hardt 1999, 2007; Hardt and Negri 2004). She works in the insurance industry. Not only does she handle numbers and charts on her computer, she deals with people and thereby has the "soft skills" to compete in the service-oriented industries of the Global North. Yet their understanding of affective labor is too limited and cannot make sense of the forms of affect that she experiences when she toils at and for an orangutan rehabilitation center—even if only for one month. Hardt and Negri's (2004) understanding of affective labor cannot make sense of the ways in which the labor she chooses to engage as a paying volunteer entails the affect of encountering a massive ape puckering his lips toward her, the intensity when he grips and rocks his cage door a few feet away from her, and the surge of sensation when she or a covolunteer can say that she felt that she was sincerely there, in the moment, far from home, doing something, feeling something (Parreñas 2012).

CHAPTER 3: FORCED COPULATION FOR CONSERVATION

1 Peter was a Bidayuh man who, like Syed at Lundu Wildlife Center, had gone into early retirement from the Forestry Department and accepted a position as a senior officer for the Forestry Corporation on a subcontractual basis.

2 Some scholars argue that the figure of the child in reproductive futurism is implicitly racialized as white (Muñoz 2007; Smith 2010).

3 I am indebted to Gayle Rubin for her explanation of the sex/gender system, in which she explains that sex is "biological, raw material" and gender is "human, social intervention" (Rubin 1975: 165).

4 Palmer et al. (2016) found that when zookeepers in New Zealand were asked about their greatest concern among their captive orangutans, they were most worried about female–female competition. Palmer's empirical research suggested that female–female competition was not an issue and that they were quite tolerant of each other. This for me serves as an example of how ape behavior is reported through culturally and socially informed perspectives—in this case a common patriarchal and perhaps setter colonial cultural trope of female competition (Haraway 1989). I further discuss the cultural frameworks that likely inform Nadim's perspective on his empirical observations of orangutans' behavior at his workplace.

5 The director of WCS had made the compelling argument directed toward the conference attendees that Ulu Sebuyau was virtually a "gold mine" for tourism since Alfred Russel Wallace theorized evolution there, contemporaneously with Charles Darwin. It could be another, more easily accessible Galápagos Islands. Such talk is reminiscent of the bioprospectors studied by Cori Hayden (2003) who considered prospecting as a potential means to profit from biodiversity.

6 The difference between subspecies may be an issue here when comparing Mentoko to Batu Wildlife Center. The subspecies *Pongo pygmaeus morio* resides in Mentoko, and females of the *morio* subspecies have the smallest ranges among females of all orangutan subspecies (Singleton et al. 2010). *Morio* also have the lowest density of all orangutan subspecies (Husson et al. 2009). Because Batu Wildlife Center is a rehabilitation center and its orangutans could have come from a variety of places before Sarawakian authorities confiscated them, only genetic tests can verify which of the subspecies populate the site. However, Sarawak and the bordering Indonesian state of West Kalimantan are populated by the subspecies *Pongo pygmaeus pygmaeus*, so the population at Batu Wildlife Center is likely the subspecies *pygmaeus*. The subspecies *pygmaeus*'s density in the wild varies from 0.35 per square kilometer at its smallest density (or about 2.75 orangutans for the area of Batu Wildlife Center) to 3.55 per square kilometer at its greatest density (or about 23.075 for the area of Batu Wildlife Center). Considering the latter figure, which is close to the figure of Batu Wildlife Center's density, requires consideration of a key hypothesis asserted by primatologists Husson et al. (2009)examining primatological empirical studies: "logging operations lead to inflated orangutan densities in neighboring, unlogged habitat" (Ancrenaz and Lackman-Ancrenaz 2004; Marshall et al. 2006; Morrogh-Bernard et al. 2003; Russon et al. 2001). Their point helps demonstrate that comparisons between "wild" orangutans in the field and rehabilitant orangutans in rehabilitation centers are not comparisons between pristine and wild nature on the one hand and unnatural, tainted culture on the other. Rather, what is observed in the wild is produced through logging and other human–animal–ecological multispecies encounters: felling large trees, lowering the availability of food sources, and the unintended growth of some plants over others are just a few outcomes of such possibilities.

7 In a restricted access test, the doors connecting cages are too small for the males to enter. Thus, female choice is evident to researchers in that the female orangutan exercises agency when using the entryway to access males. The study was conducted in 1991 at Yerkes Lab.

8 The dynamic between this couple, where the man offers up a reason for sexual violence that is rooted in sexual desire for beauty and his female companion laughs, conveys how sexuality is mediated through cultural concepts. Writing about Malay masculinity, anthropologist Michael Peletz (1996) explains that men are understood as having *gatal*, or an intrinsic itch to scratch when it comes to sexual desire. Likewise, anthropologist Suzanne Brenner (1998) explains the concept of *nakal* among Javanese men. The term references a lifelong "naughtiness" of uncontrolled sexual desire (Brenner 1998; Geertz 1961).

9 For more on male–male violence among orangutans, see Fox (2002); Mitani (1985); van Schaik et al. (2009).

10 The zookeeper I assisted at Oakland Zoo had the idea of giving birth control pills to the sun bear Ting because her performance of psychopathological behavior, specifically pacing in a tight circle when a large space was available to her, was

cyclical and thus potentially hormonal. The idea was quickly shot down by her colleagues at the zoo on similar grounds that sun bears as a species were vulnerable to extinction.

11 The ideology of liberalism is not liberal politics versus conservative politics. Rather, it is the prevailing political ideology of Britain, France, and the United States. It is the political ideology associated with nation-states, rights, and citizenship. Liberalism as a political ideology has little traction because of the long history of indefinitely deferred independence under the Raj, the colonial regime, and the postcolonial state. Liberal subjectivity, specifically citizenship, was only recently an official designation following official decolonization in 1963. However, what that citizenship gains seems to differ when it comes to social inequalities.

12 Wrangham and Peterson write, "By a logic that challenges our strongest moral principles it could pay the woman to acknowledge the rapist's power and form a relationship that, while initially repellant, she comes to accept" (1996: 142).

13 The debate about rape further highlights how scientific knowledge is produced through specific social and political contexts. See Daston (1992); Keller (1982); Latour and Woolgar (1979).

14 It is unknown whether differences between subspecies have any bearing on figures for forced copulation. The subspecies *Pongo pygmaeus pygmaeus*, located in Sarawak, has yet to be the subject of such extensive studies on mating habits as have members of *Pongo pygmaeus wurbii* in Gunung Palung of Knott's studies or of *Pongo pygmaeus morio* in Mentoko of Mitani's orangutan studies.

15 Donna Haraway (1989) detailed such underlying bias in *Primate Visions*.

16 King also fortifies her human exceptionalism through a human culture argument by citing an article that references the cultural anthropologist Peggy Reaves Sanday's (1981) study that systematically compared rape across cultures to prove the point that violence against women is not universally expressed and is always specific to particular groupings of people.

17 Crystal Feimster's (2009) startling history of rape and lynching in the American South explains the high number of reports of rape committed against enslaved women by invading Union soldiers, which were often made by white women and children. These acts were interpreted by southern whites as violations of their property.

18 For enslaved people treated as chattel, in which enslavement is matrilineal, systemic rape was almost certainly traumatic, and American society has to address, reconcile, or make reparations for such trauma, violence, and profit extraction. But for orangutans constrained in space, we cannot know what psychological trauma they experience, if any. We can only extrapolate what we read in others' faces, as Nadim does.

19 An important exception to this point is the story of Harriet Ann Jacobs (Jacobs et al. 1987), author of the autobiographical novel *Incidents in the Life of a Slave Girl*, which was originally published in 1861. Jacobs tried to evade rape by her master by forming a liaison with another white man. I am indebted to Brigitte Fielder (2013) for her research. The brutality of slavery was both permitted and

reinforced by a white supremacist sense of justice, as demonstrated by antebellum Mississippi court cases in which slave owners gained compensation from the state for executed slaves (Johnson 2013).

20 We may not know if there is a psychological impact similar to that in humans who are threatened with rape. While it would be possible to test for cortisol levels, which are hormones released in stressful situations, I very much doubt that the institution responsible for the care of this population would consent to verifying that their population is distressed. Without this kind of evidence, it could be that orangutans do not experience this as stressful. If that were the case, would these acts still constitute rape? This is an intriguing question that lacks a simple answer.

21 I think it's an interesting parallel to consider the plight of "bare sticks" in China, the population of men born during China's one-child policy. Such men are also subject to a system that superficially appears to have favored them. See Greenhalgh (2013).

22 This is where I disagree with the volunteer who I described in the introduction. While she could say that orangutans at large are like sex offenders needing rehabilitation, I think they are wrongly blamed. They are wrongly assumed to be agents of violence. Instead, I think it is more useful to approach this problem by thinking systemically and not by looking for individual culpability. To suggest an animal's culpability evokes the historical period when animals were regularly held as defendants in trials in English settler society in North America (Anderson 2004).

23 Gender is a social and cultural construct. Animals have gender when humans impose their socially and culturally informed ideologies of gender on them. See the Macmillan Interdisciplinary Handbook, *Gender: Animals* (2017).

24 That orangutan was released in March 2009, but never returned. Most feared she was dead. Others optimistically hoped she was alive with the six other orangutans released when the park was privatized.

25 Species was harder to distinguish than sex, and as anyone who has tried to sex a cat knows, it can be very difficult. The point to be taken here is that when we think about "multispecies" or "interspecies" encounters, we should never assume that "species" is a stable or easily recognizable category (Kirksey and Helmreich 2010; Livingston and Puar 2011).

26 This opens the question of whether or not dairy cows and pigs are systemically raped for the production of milk and meat. In the case of pigs, many resources have been directed toward making fertilization pleasurable for sows with the advent of artificial insemination, as Gabriel Rosenberg describes. The grounds of justice may not be on the question of whether or not they experience sexual violence, but rather on the question of how female reproductive capacities are harnessed for the sake of economic exploitation. See Bear et al. (2013).

CHAPTER 4: FINDING A LIVING

1 The pronunciation of "cari makan" is [tʃari makan]; the c is a tch sound.

2 Anatomical sex was too difficult to identify based on sight alone.

3 No one in the village in which I lived longed for the days without electricity or

for a creaky and leaky timber longhouse that would need to be replaced every few years because of rot. The conversation came up one day soon after my arrival when I had asked about the framed photos on the wall. They were of my adoptive father as a young man in traditional Iban wear. An aunt explained how a class of anthropology students from Australia had visited them for about a week in the early 1980s. They took the photos, asked questions, and came back later with the photographs. With photographs beaming with nostalgia for bark cloth, the auntie went on to explain that the photographs were from a long time ago, when they had arrived at Kampung Mohon and still lived in the old wooden longhouse. She emphasized how dramatically different their lives once had been, even within Kampung Mohon, by emphasizing the building materials of their former homes compared to their current ones. This was a situation in which nostalgia came from the outside. It was not their nostalgia that had them display such photos; it was the gift of young anthropologists in the 1980s, who with my adoptive father performed a colonial nostalgia. Yet it would be a mistake to see nostalgia as a purely colonial fantasy. Layang's, Ren's, Len's, and Julai's eagerness at taking down a bees' nest that was too close to the orangutan jungle skills training, of then being able to harvest bee larvae and enjoy the delightful taste was certainly met with a relish that can only happen after long being prevented from doing so.

4 Tim Choy (2011) sees that environmental politics in Hong Kong involving pink dolphins and residents of the ancient fishing village Tai O are structured around the idea of endangerment and that the idea of endangerment is an example of what he calls "anticipatory nostalgia," one that forecasts loss that is both spatial and temporal. While the human and animal coastal residents of Choy's study are both peripheral to Hong Kong, they lack direct contact with one another. They share a concept but not contact, and thus not the same context. This lack of contact contrasts with the case I highlight here in which loss defines the context in which humans and animals encounter one another. Direct contact and co-presence are central to being at the interface and are not the result of being marginal to a city center.

5 Camilo Jose Vergara's *American Ruins* (1999) and *New American Ghetto* (1995) offer visual examples of this. One particularly stunning example in *American Ruins* (1999: 86) is of a tree growing in the center of the second-floor reading room of the abandoned Camden Free Library. The power of that image lies in the destitution and debris in which this tree tenaciously finds a way to live. This might nourish optimism and hope in some viewers, yet I think it forces us to confront the violence of neglect.

6 Slow violence is a particularly compelling analytic for thinking about Sarawak, especially in comparison to the acute and rapid violence that has happened in the region. Sarawak has not had riots like those in neighboring West Kalimantan or in peninsular Malaysia (Peluso and Watts 2001). Additionally, Sarawakians find themselves in economic conditions less severe than some of their distant relatives on the other side of the border shared with Indonesia. Yet people in Sarawak ex-

perience themselves as less economically privileged than their fellow citizens on the peninsula. This sense of inequity informs many Sarawakians' attitudes about the federation of Malaysia, as I discuss in chapter 5.

7 Pronounced *char-ee mah-kan*.

8 "*Tuai Rumah—orang hormat, haiwan tak hormat!*"

9 As discussed in the introduction, orangutans are the only animals that are given names.

10 I think it was for my sake that Apai Julai explained that he was from Batang Ai. Much of what he had to say seemed to reveal that he was from the Third Division, perhaps upriver on the Katibas River, a tributary to the mighty Rajang River. If situated at the Katibas, it would be possible to walk to the Batang Ai region.

11 Apai Julai said, "*di batang ai, jauh di ulu, sini dekat Bandar: itu mahu pinda ke sini. Sana susa cari makan.*"

12 He was eight years older than me and left Ulu Sebuyau when he was still very young. My relationship with this van driver was amicable. Later, I was told that before he married his wife, before he went away to bejalai, he would wear women's clothing. He stopped when he came back and married his wife, who was very shrewd with money. She was "pandai" with "duit." She insisted on managing their money, invested it, and would charge people. This cleverness was a source of distrust and animosity with many in the village.

13 Many scholars of Iban studies have described bejalai as a journey taken by young men in order to gain fortune and prestige (Freeman 1970; Kedit 1993). Jensen (1974) reports that the usual length of bejalai was two to four years, followed by four to ten years.

14 The controversial Bakun Dam opened in October 2010 and flooded more than 42 km^2 and has damaged a catchment basin consisting of 1.5 million hectares of primary forest. Dam construction entailed the forced removal of approximately 10,000 forest-dwelling people of various indigenous ethnicities. Bakun is the largest dam in Southeast Asia and is scheduled to be the largest concrete-face rock-filled dam in the world. Indeed, Sarawak's mega-dam projects Bakun and Batang Ai have affected the lives of people at Lundu Wildlife Center. Layang himself, along with a five-person team from the Forestry Corporation, had to personally oversee the capture and transfer of animals on Sarawak's protected and totally protected species lists that were displaced from their habitats. For the relationship between Batang Ai Dam and Bakun Dam, see *Empty Promises, Damned Lives* (World Commission on Dams 1999).

15 I am not disclosing the real date in order to preserve the privacy of the longhouse.

16 The Cold War in Malaya and Sarawak was waged through detention and torture. British occupation during the Malaya Emergency became a model for American, British, and French warfare in the twentieth and early twenty-first centuries (Khalili 2013). The same tactics waged against civilians during the Malaya Emergency were practiced a decade later in Sarawak when the postcolonial state, supported by the British military, waged war against communist insurgents; see Yong (2013).

17 That orangutan, named Joe and then apparently renamed Bill, was transferred to Artis Zoo in Amsterdam in September 1965.

18 Horowitz (1998) briefly mentions Iban displacement in Batang Ai as a response to the communist threat. The land to which Apai Julai and his kinfolk relocated was in the Lundu area. Five years before their arrival, communists managed to kill fifteen Malaysian rangers between Lundu and the border town of Biawak (South-East Asia Treaty Organization 1973). The Research Office of the South-East Asia Treaty Organization, predecessor of the Association of Southeast Asian Nations (ASEAN), alleged that insurgents likely used guns that were homemade with bicycle inner tubes taken from three hundred bicycles reportedly stolen the previous year. Their report argues that bicycle thefts were connected to bomb construction after the importation of lead pipes, a weapon of choice for bombing, was halted in June 1971.

19 Kelvin liked to joke with me that there are more orangutans than Kelabits, so that Kelabits are indeed an endangered species. Kelabit is one of the smallest indigenous ethnic groups in Sarawak; their population is about five thousand. Following World War II, many Kelabits embraced Western education and have emerged as leaders of civil society in Sarawak and Malaysia (Bala 2007).

20 For Kelvin's take on adat and conflicts between different indigenous ethnic groups in Sarawak over land use, see Egay (2007).

21 The terms *pandai* and *pakar* do not appear in dictionaries until at least the second half of the twentieth century. Pakar is more associated with formal knowledge and is synonymous with the borrowed Arabic word *ahli*, a specialist or expert. Pandai is associated with being learned or smart. Neither has identifiable origins in *Kamus Dewan* (8th ed.), which is the standard reference for the Malay language. This dictionary describes origins as specific as the various sultanates of the Malay peninsula, so not listing an origin is surprising. Perhaps it indicates how close these ideas are culturally and linguistically to the Malay speaking world. Its lack of a definition in earlier dictionaries may simply indicate colonial attitudes toward native subjects, who were presumed to be incapable of ever being experts.

22 Lisa interprets her low-wage work at a fast food restaurant as a means of expertise, which is contrary to how most, especially scholars in the Global North, perceive such work, namely as a dead-end job (Ritzer 2015).

23 Elizabeth Povinelli's (2006) work on mutual vulnerability is particularly instructive here. Povinelli describes how a pus-filled sore on her body is the site of both Western medical interventions in the post-AIDS moment in the Global North while at the same time a site of intimacy between herself and her Beleyun kin.

24 One story that seemed to be memorable to everyone at the Rehabilitation Center was of a Chinese woman and her son. Ching had scratched and scarred the lady's legs, and so the lady demanded compensation for plastic surgery done in Japan. This story seemed to be a way of criticizing the unequal distribution of wealth experienced by people whose labor, experience, and opinions are little valued.

25 Clifford Sathers explains that Hose and McDougall have confused animism, or a

belief in the spiritual essence of all living things, with the particularities of Iban attitudes toward animals as bearers of divine signs; Sathers understands that augury is about "external guidance and a degree of confidence in those areas of life which they recognize to be particularly beset with uncertainties" (Sandin 1980: xliv).

26 Audra Simpson's (2014) "ethnographic refusal" prompts deeper thinking about this. While Hose and McDougall wish to reveal all they know about Ibans, my approach is fundamentally different. I am not trying to salvage a holistic culture here, but to fathom the transformation of relations occurring over a span of a century.

27 If the feelings my younger sisters had in regard to orangutans can be taken as an indication of what this kind of reverence might mean for younger generations, it is important to note that reverence can be paired with fear. When they would see orangutans, or see their mother holding an infant orangutan, they would describe it using the word *geli*. The Iban and Malay word *geli* conveys disgust and creepiness.

28 Angus is a traditional Scottish name. It is difficult for me to imagine anything but this being a Scotchification of a common Iban name, Anggun. However, I am not certain and thus I resort to using the name Angus.

29 The anthropology of Christianity offers similar comparison to the ways in which proselytization and conversion account for a shift from past relations to new relations (Cannell 2006; Robbins et al. 2014).

CHAPTER 5: ARRESTED AUTONOMY

1 There is some debate about the parameters of what counts as adolescence and other life stages for orangutans (Galdikas 1985; Nadler 1995).

2 The forms of arrest that confronted me while undertaking this research compel me to write this chapter in the present tense.

3 The word *bebas* is used in both Sarawakian and peninsular Malay. See Dewan Bahasa dan Pustaka (1998).

4 Arrested autonomy is similar to the arrested histories described by Carol McGranahan (2010) in her work with Tibetan refugees, whose secrets are arrested histories to be remembered and later unburied like religious texts hidden for safekeeping. The similarity lies in the idea that the arrest of the past is oriented toward a future release. However, the difference is that the arrest McGranahan describes is oriented toward a temporality and likely to be temporary. The arrest of arrested autonomy diminishes the possibility for future alternatives as the state of arrest continues indefinitely.

5 Celia Lowe (2006) and Hugh Raffles (2002) show how relations with multiple beings make up both space and place.

6 For more on "autological" subjectivities, see Povinelli (2006).

7 To be a subadult male is to have suppressed hormones that arrest the development of face flanges. They appear to be young, but are sexually mature. Thus they are categorized as subadult (Maggioncalda et al. 2002; van Schaik et al. 2009).

8 Nadim had exactly the same job as a civil servant in Sarawak's Forest Department, and was now classified as employee of its privatized branch, the Forestry Corporation. See the introduction and chapter 6 for more on the private–public partnership.

9 Field notes, June 9, 2010.

10 See Anthony Milner's *The Invention of Politics in Colonial Malaya* (1994), Anthony Reid's *Asian Freedoms* (1998), and Mary Steedly's *Rifle Reports* (2013).

11 These stereotypes emerged in the nineteenth century, when the British empire had already gained wealth from slavery and plantations, and through social and political pressure for abolition (Midgley 1992; Said 1993).

12 See Reid (1998).

13 The copy held by Yale Libraries has a handwritten inscription by one R. Gordon Gallien, Sarawak, 1918, written in both English and Jawi, or Malay written in Arabic script.

14 Wikipedia should not be idealized as "free" since a small number of editors produce most of the content on the site. See Priedhorsky et al. (2007).

15 Nadim's vision can be read as an example of Vinciane Despret's (2004) idea of anthropo-zoogenesis.

16 In respect to Sarawak's governance during the Brooke era and the Brunei era before that, Sarawak can be understood as a textbook example of James Scott's *The Art of Not Being Governed* (2009). When the Brunei sultanate nominally ruled the land, rivers connected Sarawak's many different upland ethnic groups to coastal settlements historically dominated by coastal Malay groups, who collected tribute and taxes in the name of the sultanate. However, the extensive network of rivers meant upriver people had multiple channels to acquire coastal products, such as salt and salted fish, for which they would exchange forest products such as camphor (Freeman 1957; Jawan 1994; Kedit 1980; Pringle 1970). For other examples of this long-standing history of trade, see Padoch and Peluso (1996) and Dove (2011). Once James Brooke was granted Sarawak in 1841 by the Brunei sultanate in return for suppressing a local revolt, he adopted local signs of power, such as donning the royal color yellow. He acted as sultan in an area that historically had had peripheral relations with the sultanate.

17 Sarawak under the Brooke dynasty had very few European (read: white) officers. Each of Sarawak's divisions had a European officer and district officers who reported to him (Kedit 1980). These officers were instructed to have "a mixture of kindness and freedom with severity when required without harshness or bullying" (Ward 1966: 33ff). Brooke's governance relied on native administrators, who were ethnically Malay or Melanau of the coastal regions. James Brooke and his heir Charles Brooke encouraged Chinese migration into Sarawak, where they were concentrated in urban areas. Among upriver Dayak populations in the interior, James and Charles Brooke acted as supreme war chiefs: they deployed various groups as mercenaries in wars against neighbors, which they strategically sanctioned (Jawan 1994; Pringle 1970; Wagner 1972). To the extent that headhunting could be both suppressed and sanctioned by the Brooke regime, Ibans

and other upriver people experienced autonomy because they were "protected" from changes that would "alter" their way of life, which included education, literacy, and development (Jawan 1991: 81; Pringle 1970; Reece 1982). Thus, between 1841, the time the heir to the Brunei throne ceded Sarawak to James Brooke, until 1941, the year Japan invaded and Charles Brooke's heir Charles Vyner Brooke absconded, Ibans and other upriver people were kept in a state of arrested development for the sake of what the Brooke regime felt was their own good. The paternalism the Brooke regime had for its native subjects was expressed well at the centenary celebration of Brooke rule in September 1941, when Charles Vyner Brooke officially proclaimed the Nine Cardinal Principles. The eighth principle read as follows: "That the goal of self-government shall always be kept in mind, that the people of Sarawak shall be entrusted in due course with the governance of themselves" (Yong 1998: 153). Such a promise could not be kept: three months later the Imperial Japanese Army invaded and occupied Sarawak. During the Brooke era, Sarawakians had very little opportunity to participate in larger political structures (Jawan 1991, 1993). Membership in the Council Negri, which discussed state decisions, was by invitation of the rajah; they met about once every three years (Kedit 1980). In other words, Sarawakian political autonomy was arrested.

18 While establishing native customary rights over land, it also ruled that Chinese migrants could not own land, but were able to lease it. Thus today, diasporic Chinese who were born in Sarawak and whose parents were also born in Sarawak can only lease land on which they may own a house; they cannot own the actual land.

19 Swidden agriculturalists' farming practices are pejoratively called "slash and burn." Burning bestows nitrates. Swidden agriculturalists have rotating cycles in which they usually cultivate gardens for a few years and let them go fallow in other years (Dove et al. 2005; Fried 2000; Padoch and Peluso 1996; Zerner 2000).

20 Kaur (1998) demonstrates well how these laws curtailed native rights. The rules were modeled on Mead's earlier work in the Federated Malay States and the administration of Malaya's forests. This is one way in which the peninsula and Sarawak demonstrate a history of being a center and periphery.

21 This conversation happened on June 14, 2010, between the afternoon feedings a few meters away from the construction work.

22 It often seemed to me that I as a Filipina was regarded as an intimate outsider. The state of Sarawak is too far from the Philippines to have large Filipino immigrant populations, unlike the neighboring state of Sabah. Yet the Philippines is geographically close enough to be often featured in local news media and thus was in the national imagination of Malaysia. As I have written earlier, I was racialized as local based primarily on my appearance. The fact that I was Filipina American did not seem to be as categorically important as having Filipino ethnic origins. Due to my own gendered and racialized subjectivity, I believe that my informants shared a different set of insights with me than they would have

had I been racialized in another way. I believe they shared what often could be construed as politically sensitive because they considered me as coming from the Philippines, which was much more notorious for corruption and political violence than Malaysia.

23 Field notes, June 4, 2010. When I shared this story with Amali Ibrahim, author of the forthcoming *Improvising Islam*, he suspected that this exemplifies a new environmentalist interpretation of Islam.

24 Examples of this genre include Baring-Gould and Bampfylde's *A History of Sarawak* (1909); Robert Payne's *The White Rajahs of Sarawak* (1986); Tom Harrisson's *A World Within* (1959); *Taib: Hero of Development* (Taib Mahmud and Gaya Media Sdn. Bhd. 1996); *Footprints in Sarawak* (Ong 1998); and Stephen Yong's *A Life Twice Lived* (1998).

25 Field notes, November 12, 2009.

26 As Stephen Yong (1998: 193) wrote, "many native ethnic groups felt that they were bumiputera in name only and were 'second-class' bumiputera."

27 See Freeman (1957); Hose (1926); Wallace (1986); and Sandin (1980). Jayum Jawan (1994) refutes this pervasive description in *Iban Politics and Economic Development*; he states that wealth varied between individuals and thus created a hierarchy.

28 This sparked Sarawak's first political movement, spearheaded by Malay civil servants of the Raj. Anticessonists protested for the restitution of Brooke rule, supporting Charles Vyner Brooke's nephew Anthony Brooke's potential claim to the throne in the hope that Sarawak would retain its independence. Those active in the movement were mostly urban Malays, a group that filled an administrative niche in the Brooke regime. Their actions culminated in the December 1949 assassination of the second British governor of Sarawak, which led to the execution of two of the movement's leaders. The movement was swiftly suppressed by the British. See Stephen Yong's autobiography *A Life Twice Lived* (1998); Kedit (1980). Reflecting the Brooke administration's autocracy, Charles Vyner Brooke made this proclamation without having consulted his brother, Bertram Brooke, the Tuan Muda with whom he technically shared power, nor the named heir Anthony Brooke, who held the title Rajah Muda. See Payne (1986).

29 According to the June 1962 census, Sarawak's population was 776,990; 51 percent were Dayak, mostly Iban, 31.5 percent were Chinese, and 17.5 percent were Malay. See Ongkili (1967).

30 Layang had very specific ideas about what should happen at Lundu Wildlife Center in regard to three semi-wild orangutans. The corporation was entirely against his plans. Layang, however, was bold: he found a way to defy the corporation's control. Layang "forgot" to lock the cage of Ching, a female orangutan famous for biting local women.

31 The Iban speakers with whom I lived and worked in their casual speech would call orangutans "orang," which in Malay literally means "people" or "persons." I do not believe that people who speak about orangutans as persons literally see orangutans as persons. Rather, I think it is indicative of the colloquial usage of

Malay as spoken by Ibans. In Malay, direct and indirect objects are linguistically organized, counted, and referred to and marked according to the type of object they are categorized as. When objectified through language, sentient beings in Malay are either "orang" (persons) or "ekor" (tails). In Iban and its dialects, there is only one term for both humans and animals: *iko*, which means tail. This linguistic identification, perhaps, is indicative of pagan Iban belief systems that have since lost their appeal in the face of modernization through Christian conversion and wage labor. A musing 'Kak shared with me and that I describe in chapter 4 indicates the transformations of human–animal relations in Iban contexts: "The old people are capable of saying that bird means this, that bird means that. They're all just birds to me!"

32 This village consisted of economic migrants from a variety of upland, forested interiors. People of Iban ethnicity settled here from Ulu Sebuyau, Batang Ai, Lubok Antu, and Betong—all of which are Iban communities with distinctly different lexicons. All came to this lowland place closer to the capital city in order to participate in the wage economy. I explain this in chapter 4.

33 I discuss wage labor as a means of livelihood following displacement in chapter 4.

34 Similarly, before the security guard ultimately was fired for drinking too much on the job, his twelve-year-old daughter would cover for him and do his night shift from 7 PM until 7 AM. When asked where her father was, she would say that he was either bathing or at the toilet.

35 We spoke in Malay, but I am reproducing our conversation here in English.

36 Interview, March 5, 2010.

37 See Jayum Jawan (1994) and Peter Kedit (1980) on wage earning among Ibans in Sarawak. 'Kak eventually received a raise, as did her husband Ren. After an absence of six years, I was shocked to find that they had given up on rice cultivation and maintaining their small garden plot. They gave up on their food sovereignty and now must buy all their food, which is quite precarious as the Malaysian federal state has imposed a general sales tax, even on uncooked food, and as the ringgit decreases in value at the time of writing.

38 Carolyn Merchant (1980) also discusses the mechanical order in the work of Hobbes and his contemporaries.

39 In Varela's work, the abstract idea of autonomy is physically evidenced through autopoiesis in living systems, in which living systems "produce their own identity; they distinguish themselves from their background" (1979: 13). Here, the background is the environment and not genealogy. And production is through dynamic relations between components, not just static components standing in relation to each other. And individuality in living systems, whether referencing a cell in a sampling of cells or individuals in a social unit, has boundaries between itself and all that is at its interface.

40 On different occasions and under different circumstances that I experienced, men who were in their late fifties and early sixties in the 2010s, and who were either teenagers or young men at the time of official decolonization, spoke wist-

fully of their youth when Sarawakian Malays would celebrate Hari Raya with langkau and tuak—Sarawak rice whiskey and wine—which today is simply haram, that is, forbidden for Malays for whom their racial identities are tied to Islam. When they shared this recollection with me, they would always explain that this is positively shocking for anyone unfamiliar with Sarawak, especially Malays from the peninsula.

41 Risks for Nadim and 'Kak could include the risk of violating ritual prohibitions or the risk of misunderstanding and miscommunication.

42 This for me evokes Jacques Derrida's (1977) idea of the trace that Gayatri Spivak explains in her preface to her translation of that work.

CHAPTER 6: HOSPICE FOR A DYING SPECIES

1 Accredited zoos in the Association of Zoos and Aquariums, for instance, meet standards set by the organization in regard to animal welfare, management, veterinary care, and safety. Accredited members must pass review every five years. These standards are reviewed and revised annually by the accreditation board. The standards are high, and less than 10 percent of animal exhibitors licensed by the U.S. Department of Agriculture are accredited by the AZA, which is the leading standard-bearer of animal management in North America. Aware of the standards of zoos, a volunteer once said to me that Lundu Wildlife Center was like a second-rate zoo.

2 The infant was conceived at the time of the original photograph in figure 3.1 was taken, in 2006. The young age of the mother was not taken into account as a reason for their deaths. Rather, it was believed that the fact of captivity killed them along with a diet of cold foods. Eating only warm food is common postnatal human care throughout Borneo.

3 Field notes, November 12, 2009.

4 His suppliers included longhouse dwellers, including 'Kak and her neighbors, before they sold their plots to one of their neighbors, who has since converted them to an oil palm plantation. 'Kak did not sell hers, but instead poured concrete over it, and now her son has a welding workshop on the grounds.

5 The list of fatalities also includes a proboscis monkey that was surrendered in October 2008 after having been fed a sweet cake and who quickly died from intestinal problems. Two volunteers had the grim task of having to bury the almost twenty-five-kilo (fifty-pound) monkey behind the quarantine.

6 A particularly poignant case was of a male proboscis monkey who had been given a sweet baked cake to eat. Their stomachs cannot digest sweets. The proboscis looked uncomfortable; it was sitting up, but slumped over and holding still. The proboscis shortly died.

7 Lin's other languages were Foochow, Mandarin, and Malay—both official Bahasa Malaysia and Sarawakian Malay.

8 Commercial volunteers through ENGAGE built the ranger station.

9 The law creating the Forestry Corporation specified five revenue streams: "The Fund of the Corporation shall consist of—(a) all moneys received by the Corpo-

ration for services rendered by the Corporation to the Government as its agent or for services rendered by the Corporation to any person; (b) all moneys received by the Corporation by way of grants from the Government; (c) all moneys derived from the disposal, lease, or hire of, or any other dealing with, any property vested in or acquired by the Corporation; (d) all moneys derived as income from investment by the Corporation; (e) all moneys borrowed by the Corporation under this Ordinance; and (f) all other moneys lawfully received by the Corporation."

10 The Forestry Corporation's instrumentalization was not just its potential to act as creditor and debtor. The last words of the law establish the corporation's power to pay both the board and anyone they invite at the discretion of the minister of natural resource planning and development.

11 Documents in the orangutan file of the Sarawak Museum archives suggest this. The resolution was to have multilingual signs indicating that entering certain areas of the park would be at one's own risk.

12 A comparison with the city of Howrah outside of Kolkata would be fruitful since both Sandakan and Howrah were hinterland industrial ports that were in some senses preindustrial and are now postindustrial (Majumder 2013).

13 The concept of running amok has captured the imagination of anthropologists and psychiatrists alike. The phenomenon of *amok* is "an outburst of murderous aggression found almost exclusively among adult males in the Malay world," specifically present-day Malaysia, Indonesia, and the Philippines (Schmidt et al. 1977: 265; see also Good and Good 2001; Kon 1994; Spores 1988). Running amok sounds akin to "atavism," or degeneration into a base animal (Pick 1989). However, this idea does not stand up to the question of what it would mean that an ape is capable of going amok. It seems to suggest that dangerous masculinity conveyed by the idea of amok is regardless of humanity or animality.

CONCLUSION: LIVING AND DYING TOGETHER

1 For thinking through elemental time scales and wind, see Jerry Zee's (2017) work on sand.

2 Conservation International issues this list in consultation with attendees at the biennial International Primatological Society congress. I attended the 2016 meeting. Russell Mittermeier, president of CI, tasked each continent to prioritize a list of a specific number. Because Madagascar is exceptional in both its species diversity and its conservation threats, it is considered its own continent. The decision to add ringtail lemurs and aye-ayes on the list was controversial at the meeting. Controversy was expressed with grumbles and murmurs during the meeting and unabashed verbal complaint in the aftermath of the meeting.

3 They did not transfer totally protected species that they thought could escape on their own, namely wild boar and crocodilian species.

4 The report is dedicated to the five individuals who died in the aftermath of the project. The number of casualties differs from the initial news stories, which indicated that ten had passed away.

5 Evans-Pritchard (1937) explains the workings of witchcraft through a case in which someone died when he was sitting under a granary. He pointed out that termites had eaten through the post supporting the granary. That was obvious to everyone, but that was not a sufficient explanation for why the granary fell when it did. Witchcraft was considered the reasonable explanation for that. See Steedly (1993) and Klima (2002) for communication with the dead following violent massacres in Southeast Asia.

6 The bacterium responsible for melioidosis could serve as a case of Heather Paxson's (2013) "microbiopolitics," where microbes get categorized and judged as either bad or good and thus deserving to respectively diminish or flourish, despite the fact that life as we know it stems from microbial relations.

7 Sarawak guards against rabies by having a long quarantine period. However, quarantine did not protect Sarawak from its first ever rabies outbreak in 2017.

8 *Bunsuh* is a kin term in Iban for youngest child. It is related to the Tagalog *bunsoh* and Malay *bongsu*.

9 By 2016, 'Kak and her family had given up on rice cultivation and even their small garden where they grew pineapple, durian, and sweet potato. Sweet potato plants had offered them a regenerative supply of leafy greens in the past. White rice that they purchase has replaced their mixed red and white variety. They say that kampung rice is still delicious, as is kampung durian, but they do not have the time to go to the rice padi. By this time, 'Kak and her husband Ren were each making 1,000 RM, although the imposition of the general sales tax, which is even applied to groceries, cuts into their earnings. To me, their quitting agriculture and giving up on food sovereignty indicates a confidence that they will always be in a position to buy food. It also conveys their buy-in to the idea that Sarawak is shifting from an agrarian and resource-producing state to a service economy. Apai Len continues to cultivate his own rice for his family. He prefers a black variety.

10 There are many interesting points of comparison and contrast with Anna Tsing's research here, and I suspect the differences are not just cultural variation, but different historical circumstances. Honey trees have a similar classification for Ibans and Meratus Dayaks, but are known by different terms. For Ibans, the term is *tapang*, and for Meratus Dayaks, the term is *linuh*. The one carrying out the raid gains the honor of distributing the spoils in both groups. While in Anna Tsing's experiences, the honeycomb and brood comb were distributed among the community. In my singular—perhaps anecdotal—experience, they were only distributed among the households that participated in the *napang* or honey hunt itself. There may be a number of reasons for this. Perhaps it was simply scarcity—the fact that there was only one honey tree in the area—whereas in the context that Anna Tsing described, there were many trees. Perhaps also it was the borderline illegality of their actions and a way to limit risk. It would also be plausible to think that the conditions of wage labor have changed social relations between longhouse residents so that sharing is no longer as common as it had been in the recent past.

11 His son was not related by blood; however, bee sting allergies are a known and common phenomenon in Sarawak.

12 I intentionally leave open the question of how to reconcile the problem of frequent forced copulation with decolonization. There is no easy answer, but I think that it will take both the desire to think with orangutan perspectives and a depth of imagination to come up with interventions that could foster freedom without fear. Might it be possible to better facilitate female choice in wildlife centers by devising a large cage with multiple small entrances? Or could female choice be facilitated by hosting female and male populations at different wildlife centers? If decolonization is, as I argue, an experiment after colonialism ends, then we must be open to the possibility of uncertainty and maybe even failure. I place hope in the idea that experiments lead to new experiments, that decolonization is an ongoing project without a settled answer.

REFERENCES

Abdulgani, Roeslan. 1955. *Asia-Africa Speaks from Bandung*. Djakarta: Ministry of Foreign Affairs, Republic of Indonesia.

Adams, Vincanne. 2016. *Metrics: What Counts in Global Health, Critical Global Health—Evidence, Efficacy, Ethnography*. Durham, NC: Duke University Press.

Agamben, Giorgio. 1998. *Homo Sacer: Sovereign Power and Bare Life*. Stanford, CA: Stanford University Press.

Agard-Jones, Vanessa. 2013. "Bodies in the System." *Small Axe: A Caribbean Journal of Criticism* 17 (3): 182–92.

Agrawal, Arun. 2005. *Environmentality: Technologies of Government and the Making of Subjects*. Durham, NC: Duke University Press.

Ahuja, Neel. 2016. *Bioinsecurities: Disease Interventions, Empire, and the Government of Species*. Durham, NC: Duke University Press.

Alexander, M. Jacqui, and Chandra Talpade Mohanty. 1997. *Feminist Genealogies, Colonial Legacies, Democratic Futures*. New York: Routledge.

Alexander, Michelle. 2010. *The New Jim Crow: Mass Incarceration in the Age of Colorblindness*. New York: New Press.

Alger, Janet M., and Steven F. Alger. 1999. "Cat Culture, Human Culture: An Ethnographic Study of a Cat Shelter." *Society and Animals* 7 (3): 199–218.

Allen, Jafari Sinclaire, and Ryan Cecil Jobson. 2016. "The Decolonizing Generation:

(Race and) Theory in Anthropology since the Eighties." *Current Anthropology* 57 (2): 129–48.

Allen, Richard B. 2014. "Slaves, Convicts, Abolitionism and the Global Origins of the Post-Emancipation Indentured Labor System." *Slavery and Abolition* 35 (2): 328–48.

Allison, Anne. 2013. *Precarious Japan*. Durham, NC: Duke University Press.

Ancrenaz, M., and I. Lackman-Ancrenaz. 2004. *Orangutan Status in Sabah: Distribution and Population Size. Hutan-SWD Report*. Kota Kinabalu, Malaysia: Hutan.

Anderson, Benedict R. O'G. 1998. *The Spectre of Comparisons: Nationalism, Southeast Asia, and the World*. London: Verso.

Anderson, Virginia DeJohn. 2004. *Creatures of Empire: How Domestic Animals Transformed Early America*. New York: Oxford University Press.

Anderson, Warwick. 2006. *Colonial Pathologies: American Tropical Medicine, Race, and Hygiene in the Philippines*. Durham, NC: Duke University Press.

Anderson, Warwick, and Vincanne Adams. 2008. "Pramoedya's Chickens: Postcolonial Studies of Technoscience." In *The Handbook of Science and Technology Studies*, edited by Edward J. Hackett, Olga Amsterdamska, Michael Lynch, and Judy Wajcman, 181–204. Cambridge, MA: MIT Press.

Andrews, Thomas G. 2008. *Killing for Coal: America's Deadliest Labor War*. Cambridge, MA: Harvard University Press.

Armitage, David. 2000. *The Ideological Origins of the British Empire*. Cambridge: Cambridge University Press.

Arora, N., A. Nater, C. P. van Schaik, E. P. Willems, M. A. van Noordwijk, B. Goossens, N. Morf, M. Bastian, C. Knott, H. Morrogh-Bernard, N. Kuze, T. Kanamori, J. Pamungkas, D. Perwitasari-Farajallah, E. Verschoor, K. Warren, and M. Krützen. 2010. "Effects of Pleistocene Glaciations and Rivers on the Population Structure of Bornean Orangutans (*Pongo pygmaeus*)." *Proceedings of the National Academy of Sciences of the United States of America* 107 (50): 21376–81.

Arvin, Maile, Eve Tuck, and Angie Morrill. 2013. "Decolonizing Feminism: Challenging Connections between Settler Colonialism and Heteropatriarchy." *Feminist Formations* 25 (1): 8–34.

Ashcroft, Bill, Gareth Griffiths, and Helen Tiffin. 1989. *The Empire Writes Back: Theory and Practice in Post-Colonial Literatures*, 2nd ed. London: Routledge.

Baay, Reggie. 2015. *Daar werd wat gruwelijks verricht: slavernij in Nederlands- Indië*. Amsterdam: Athenaeum-Polak and Van Gennep.

Baer, A. 2008. "Orangutan: History, Myth, Oblivion." *Sarawak Museum Journal* 65 (86): 339–54.

Bala, Poline. 2007. "From Highlands to Lowlands: Kelabit Women and Their Migrant Daughters." In *Village Mothers, City Daughters: Women and Urbanization in Sarawak*, edited by Hew Cheng Sim, 120–39. Singapore: Institute of Southeast Asian Studies.

Baldry, Eileen, David Brown, Mark Brown, Chris Cunneen, Melanie Schwartz, and Alex Steel. 2011. "Imprisoning Rationalities." *Australian and New Zealand Journal of Criminology* 44 (1): 24–40. doi: 10.1177/0004865810393112.

Baptist, Edward E. 2014. *The Half Has Never Been Told: Slavery and the Making of American Capitalism*. New York: Basic Books.
Barad, Karen Michelle. 2007. *Meeting the Universe Halfway: Quantum Physics and the Entanglement of Matter and Meaning*. Durham, NC: Duke University Press.
Baring-Gould, S., and C. A. Bampfylde. 1909. *A History of Sarawak under Its Two White Rajahs 1839–1908*. London: Southeran.
Barua, Maan. 2011. "Mobilizing Metaphors: The Popular Use of Keystone, Flagship and Umbrella Species Concepts." *Biodiversity and Conservation* 20 (7): 1427–40.
Basu, Srimati. 2011. "Sexual Property: Staging Rape and Marriage in Indian Law and Feminist Theory." *Feminist Studies* 37 (1): 185–211.
Bear, Laura, Karen Ho, Anna Tsing, and Sylvia Yanagisako. 2015. "Gens: A Feminist Manifesto for the Study of Capitalism." Theorizing the Contemporary, Cultural Anthropology website, March 30, 2015. http://www.culanth.org/fieldsights/652-gens-a-feminist-manifesto-for-the-study-of-capitalism.
Beck, Benjamin, et al. 2007. *Best Practice Guidelines for the Reintroduction of Apes*. Gland, Switzerland: World Conservation Union (IUCN).
Beder, Sharon. 2002. *Global Spin: The Corporate Assault on Environmentalism*. Dartington, UK: Green Books.
Bellwood, Peter S., James J. Fox, and D. T. Tryon, eds. 1995. *The Austronesians: Historical and Comparative Perspectives*. Canberra: Department of Anthropology as part of the Comparative Austronesian Project, Research School of Pacific and Asian Studies, Australian National University.
Bentham, Jeremy, J. H. Burns, and H. L. A. Hart. 1996. *The Collected Works of Jeremy Bentham: An Introduction to the Principles of Morals and Legislation*. Oxford: Oxford University Press.
Berger, John. 1980. *About Looking*. New York: Pantheon Books.
Berlant, Lauren Gail. 2011. *Cruel Optimism*. Durham, NC: Duke University Press.
Besky, Sarah. 2014. *The Darjeeling Distinction: Labor and Justice on Fair-Trade Tea Plantations in India*. Berkeley: University of California Press.
Bessire, Lucas, and David Bond. 2014. "Ontological Anthropology and the Deferral of Critique." *American Ethnologist* 41 (3): 440–56. doi: 10.1111/amet.12083.
Bhabha, Homi K. 1994. *The Location of Culture*. London: Routledge.
Biehl, João Guilherme. 2005. *Vita: Life in a Zone of Social Abandonment*. Berkeley: University of California Press.
Blanchette, Alex. 2015. "Herding Species: Biosecurity, Posthuman Labor, and the American Industrial Pig." *Cultural Anthropology* 30 (4): 640–69.
Blussé, Leonard, and F. S. Gaastra. 1998. *On the Eighteenth Century as a Category of Asian History: Van Leur in Retrospect*. Brookfield, VT: Ashgate.
Boellstorff, Tom. 2004. "The Emergence of Political Homophobia in Indonesia: Masculinity and National Belonging." *Ethnos* 69 (4): 465–86. doi: 10.1080/0014184042000302308.
Bohme, Susanna Rankin. 2015. *Toxic Injustice: A Transnational History of Exposure and Struggle*. Berkeley: University of California Press.

Bonilla, Yarimar. 2015. *Non-Sovereign Futures: French Caribbean Politics in the Wake of Disenchantment*. Chicago: University of Chicago Press.
Bontius, Jacobus. [1642] 1931. *On Tropical Medicine*. Amsterdam: Nederlandsch Tiijdschrift voor Geneeskunde.
Bourke, Joanna. 2007. *Rape: A History from 1860 to the Present Day*. London: Virago.
Braitman, Laurel. 2014. *Animal Madness: How Anxious Dogs, Compulsive Parrots, and Elephants in Recovery Help Us Understand Ourselves*. New York: Simon and Schuster.
Braudel, Fernand. 1958. "Histoire et Sciences sociales: La longue durée." *Annales: Histoire, Sciences Sociales* 13 (4): 725–53.
Braverman, Irus. 2015. *Wild Life: The Institution of Nature*. Stanford, CA: Stanford University Press.
Brenner, Suzanne April. 1998. *The Domestication of Desire: Women, Wealth, and Modernity in Java*. Princeton, NJ: Princeton University Press.
Brooke, James, and Henry Drummond. 1853. *A Vindication of His Character and Proceedings in Reply to the Statements Privately Printed and Circulated by Joseph Hume, Esq. M.P.: Addressed to Henry Drummond, Esq. M.P.* London: J. Ridgway.
Brosius, J. Peter. 1999. "Green Dots, Pink Hearts: Displacing Politics from the Malaysian Rain Forest." *American Anthropologist* 101 (1): 36–57. doi: 10.1525/aa.1999.101.1.36.
Burton, Antoinette M. 1994. *Burdens of History: British Feminists, Indian Women, and Imperial Culture, 1865–1915*. Chapel Hill: University of North Carolina Press.
Butler, Judith. 2004. *Precarious Life*. London: Verso.
Callon, Michel. 1986. "Some Elements of a Sociology of Translation: Domestication of the Scallops and the Fishermen of St Brieuc Bay." In *Power, Action and Belief: A New Sociology of Knowledge*, edited by John Law, 196–233. London: Routledge and Kegan Paul.
Cammack, Paul. 2012. "The G20, the Crisis, and the Rise of Global Developmental Liberalism." *Third World Quarterly* 33 (1): 1–16.
Candea, Matei. 2010. "'I Fell in Love with Carlos the Meerkat': Engagement and Detachment in Human–Animal Relations." *American Ethnologist* 37 (2): 241–58. doi: 10.1111/j.1548-1425.2010.01253.x.
Cannell, Fenella. 2006. *The Anthropology of Christianity*. Durham, NC: Duke University Press.
Card, Claudia. 1996. "Rape as a Weapon of War." *Hypatia* 11 (4): 5–18.
Cassidy, Rebecca, and Molly H. Mullin, eds. 2007. *Where the Wild Things Are Now: Domestication Reconsidered*. Oxford: Berg.
Caswell, J. L., and K. J. Williams. 2007. "Respiratory System." In *Jubb, Kennedy and Palmer's Pathology of Domestic Animals*, edited by M. G. Maxie, 567. Philadelphia: Saunders.
Caton, Steven C. 1987. "Contributions of Roman Jakobson." *Annual Review of Anthropology* 16:223–60.

Cattelino, Jessica R. 2008. *High Stakes: Florida Seminole Gaming and Sovereignty.* Durham, NC: Duke University Press.

Cawthon Lang, Kristina. 2005. "Primate Factsheets: Orangutan (*Pongo*)." Accessed September 1, 2017. http://pin.primate.wisc.edu/factsheets/entry/orangutan/behav.

Chan, Barney. 2008. *Institutional Restructuring in Sarawak, Malaysia.* Bangkok: Food and Agriculture Organization of the United Nations, Regional Office for Asia and the Pacific.

Chan, Zora. 2009. "Higher Orangutan Population Expected in S'wak." *Borneo Post,* November 3.

Cheah, Pheng. 2003. *Spectral Nationality: Passages of Freedom from Kant to Postcolonial Literatures of Liberation.* New York: Columbia University Press.

Chen, Mel Y. 2012. *Animacies: Biopolitics, Racial Mattering, and Queer Affect.* Durham, NC: Duke University Press.

Cheng, Henry. 2009. "State Govt Targets to Increase Number of Orangutans to 4,000: Len." *Borneo Post,* November 3.

Chin, James U. H., and Jayl Langub, eds. 2007. *Reminiscences: Recollections of Sarawak Administrative Service Officers.* Subang Jaya, Selangor, Malaysia: Pelanduk Publications.

Chow, Joanne. 1996. "Feeding Behaviour in Relation to Tree Destruction by Rehabilitant Orang Utans." Master's thesis, St. Hugh's College, Oxford University.

Choy, Tim. 2011. *Ecologies of Comparison: An Ethnography of Endangerment in Hong Kong.* Durham, NC: Duke University Press.

Chua, Liana. 2017. "Animal Internet Stardom." In *Gender: Animals,* edited by Juno Salazar Parreñas. Farmington Hills, MI: Macmillan Reference USA.

Cohen, Joshua. 1986. "Reflections on Rousseau: Autonomy and Democracy." *Philosophy and Public Affairs* 15 (3): 275–97.

Cole, F. J. 1944. *A History of Comparative Anatomy from Aristotle to the Eighteenth Century.* London: Macmillan.

Collard, Rosemary-Claire, Jessica Dempsey, and Juanita Sundberg. 2015. "A Manifesto for Abundant Futures." *Annals of the Association of American Geographers* 105 (2): 322–30.

Comaroff, Jean. 1985. *Body of Power Spirit of Resistance: The Culture and History of a South African People.* Chicago: University of Chicago Press.

Cooke, Fadzilah Majid. 2006. *State, Communities and Forests in Contemporary Borneo.* Canberra: Australian National University E Press.

Cooper, L. T., Jr. 2009. "Myocarditis." *The New England journal of medicine* 360 (15):1526–38.

Cramb, R. A. 1979. *A Farm Plan for the Proposed Batang Ai Resettlement Area.* Kuching, Malaysia: Department of Agriculture, Planning Division.

Crawfurd, John. 1852. *A Grammar and Dictionary of the Malay Language, with a Preliminary Dissertation.* London: Smith, Elder and Co.

Cribb, R. B., Helen Gilbert, and Helen Tiffin. 2014. *Wild Man from Borneo: A Cultural History of the Orangutan.* Honolulu: University of Hawai'i Press.

Cronon, William. 1996. *Uncommon Ground: Rethinking the Human Place in Nature*. New York: W. W. Norton.
Cruickshank, J. C. 1949. "Failure of Aureomycin in Experimental Melioidosis." *British Medical Journal* 2 (4624): 410–11.
Dance, D. A. B. 1999. "Melioidosis." In *Tropical Infectious Diseases: Principles, Pathogens and Practice*, edited by R. L. Geurrant, D. H. Walker, and P. F. Weller, 430–37. Philadelphia: Churchill Livingstone.
Das, Veena. 2007. *Life and Words: Violence and the Descent into the Ordinary*. Berkeley: University of California Press.
Daston, Lorraine. 1992. "Objectivity and the Escape from Perspective." *Social Studies of Science* 22 (4): 597–618. doi: 10.1177/030631292022004002.
Dave, Naisargi N. 2014. "Witness: Humans, Animals, and the Politics of Becoming." *Cultural Anthropology* 29 (3): 433–56. doi: 10.14506/ca29.3.01.
Davidson, Desmond. 2014. "Taib Mahmud tidak patut jadi gabenor hingga diyakini bebas rasuah, kata pengkritik." *Malaysian Insider*, February 20. http://www.themalaysianinsider.com/bahasa/article/taib-mahmud-tidak-patut-jadi-gabenor-hingga-diyakini-bebas-rasuah-kata-peng—sthash.OtNcFh9b.dpuf.
Davis, Angela Y. 1981. *Women, Race, and Class*. New York: Random House.
Davis, Angela Y. 2003. *Are Prisons Obsolete?* New York: Seven Stories Press.
Davis, Angela Y. 2012. *The Meaning of Freedom: And Other Difficult Dialogues*. San Francisco: City Lights Books.
de la Bellacasa, María Puig. 2012. "Nothing Comes without Its World: Thinking with Care." *Sociological Review* 60 (2): 197–216.
de la Cadena, Marisol. 2010. "Indigenous Cosmopolitics in the Andes: Conceptual Reflections beyond 'Politics.'" *Cultural Anthropology* 25 (2): 334–70. doi: 10.1111/j.1548-1360.2010.01061.x.
De Leon, Jason, and Michael Wells. 2015. *The Land of Open Graves: Living and Dying on the Migrant Trail*. Berkeley: University of California Press.
D'Emilio, John, and Estelle B. Freedman. 1988. *Intimate Matters: A History of Sexuality in America*. New York: Harper and Row.
Department of Justice. 2016. "United States Seeks to Recover More Than $1 Billion Obtained from Corruption Involving Malaysian Sovereign Wealth Fund." Press release, July 20. https://www.justice.gov/opa/pr/united-states-seeks-recover-more-1-billion-obtained-corruption-involving-malaysian-sovereign.
Derrida, Jacques. [1974] 1997. *On Grammatology*, corrected ed. Translated by Gayatri Chakravorty Spivak. Baltimore: Johns Hopkins University Press.
Despret, Vinciane. 2004. "The Body We Care For: Figures of Anthropo-Zoo--Genesis." *Body and Society* 10 (2–3): 111–34.
de Waal, F. B. 1995. "Bonobo Sex and Society." *Scientific American* 272 (3): 82–88.
Dewan Bahasa dan Pustaka. 1998. *Daftar kata dialek Melayu Sarawak: Dialek Melayu Sarawak-bahasa Malaysia, bahasa Malaysia-dialek Melayu Sarawak*, 2nd ed. Kuala Lumpur: Dewan Bahasa Pustaka.
Dimbab, Ngidang, Sanggin Spencer Empading, and Saleh Robert Menua, eds.

2000. *Iban Culture and Development in the New Reality: Iban Cultural Seminar 1998*. Kuching, Malaysia: Dayak Cultural Foundation.
Domínguez, Gabriel. 2015. "Malaysian PM Facing Biggest Crisis Yet." *Deutsche Welle*, July 14, 2015. http://dw.com/p/1FxtP.
Dove, Michael. 2011. *Banana Tree at the Gate: A History of Marginal Peoples and Global Markets in Borneo*. New Haven, CT: Yale University Press.
Dove, Michael, Percy E. Sajise, and Amity Appell Doolittle. 2005. *Conserving Nature in Culture: Case Studies from Southeast Asia*. New Haven, CT: Yale Southeast Asia Studies.
Drea, Christine M., and Kim Wallen. 2003. "Female Sexuality and the Myth of Male Control." In *Evolution, Gender, and Rape*, edited by Cheryl Brown Travis. Cambridge, MA: MIT Press.
Edelman, Lee. 2004. *No Future: Queer Theory and the Death Drive*. Durham, NC: Duke University Press.
Egay, Kelvin. 2007. "Matter of Access, Not Rights: Indigenous Peoples, External Institutions and Their Squabbles in Mid-Tinjar River, Sarawak." Working Paper Series, University of Malaysia Sarawak Faculty of Social Sciences. Kuching, Malaysia: Heng Sing Brothers.
Eng, David L., and David Kazanjian. 2003. *Loss: The Politics of Mourning*. Berkeley: University of California Press.
Errington, Frederick Karl, Fuzikura Tatsuro, and Deborah B. Gewertz. 2013. *The Noodle Narratives: The Global Rise of an Industrial Food into the Twenty-First Century*. Berkeley: University of California Press.
Esquibel, Catriona Rueda, and Luz Calvo. 2013. "Decolonize Your Diet." *Nineteen Sixty Nine: An Ethnic Studies Journal* 2 (1): 1–5.
Evans-Pritchard, E. E. 1937. *Witchcraft, Oracles and Magic among the Azande*. Oxford: Clarendon Press.
Fabian, Ann. 2013. "Charming Toads." *Michigan Quarterly Review* 52 (1). http://hdl.handle.net/2027/spo.act2080.0052.102.
Fanon, Frantz. 1965. *The Wretched of the Earth*. New York: Grove Press.
Fanon, Frantz. 1967. *A Dying Colonialism*. New York: Grove Press.
Fanon, Frantz. 2008. *Black Skin, White Masks*. New York: Grove Press.
Federal Bureau of Investigation. 2016. "U.S. Seeks to Recover $1 Billion in Largest Kleptocracy Case to Date." Accessed September 19, 2016. https://www.fbi.gov/news/stories/us-seeks-to-recover-1-billion-in-largest-kleptocracy-case-to-date.
Federici, Silvia. 2012. *Revolution at Point Zero: Housework, Reproduction, and Feminist Struggle*. Oakland, CA: PM Press.
Feimster, Crystal Nicole. 2009. *Southern Horrors: Women and the Politics of Rape and Lynching*. Cambridge, MA: Harvard University Press.
Feld, Steven, and Keith H. Basso. 1996. *Senses of Place*. Santa Fe, NM: School of American Research Press.
Fielder, Brigitte Nicole. 2013. "Animal Humanism: Race, Species, and Affective Kinship in Nineteenth-Century Abolitionism." *American Quarterly* 65 (3): 487–514.

Fielder, Brigitte Nicole. 2017. "Chattel Slavery." In *Gender: Animals*, edited by Juno Salazar Parreñas. Farmington Hills, MI: Macmillan Reference USA.

Fisher, James, Noel Simon, Jack Vincent, and Survival Service Commission of the International Union for Conservation of Nature and Natural Resources. 1969. *Wildlife in Danger.* New York: Viking Press.

Fiskesjö, Magnus. 2017. "China's Animal Neighbors." In *The Art of Neighbouring: Making Relations across China's Borders*, edited by Martin Saxer and Juan Xhang. Amsterdam: Amsterdam University Press.

Foitová, Ivona, Boužena Koubková, V. Barus, and W. Nurcahyo. 2008. "Presence and Species Identification of the Gapeworm *Mammomonogamus laryngeus* (Raillet, 1899) (Syngamidae: Nematoda) in a Semi-Wild Population of Sumatran Orangutan (*Pongo abelii*) in Indonesia." *Research in Veterinary Science* 84 (2): 232–36.

Foucault, Michel. 1978. *The History of Sexuality.* Translated by Robert Hurley. New York: Pantheon.

Foucault, Michel. 2007. *Security, Territory, Population: Lectures at the Collège de France, 1977–78.* Translated by Michel Senellart, François Ewald, and Alessandro Fontana. Basingstoke, UK: Palgrave Macmillan.

Fox, Elizabeth A. 2002. "Female Tactics to Reduce Sexual Harassment in the Sumatran Orangutan (*Pongo pygmaeus abelii*)." *Behavioral Ecology and Sociobiology* 52 (2): 93–101.

Freccero, Carla. 2011. "Carnivorous Virility; or, Becoming-Dog." *Social Text* 29 (1): 177–95.

Freedman, Estelle B. 2013. *Redefining Rape: Sexual Violence in the Era of Suffrage and Segregation.* Cambridge, MA: Harvard University Press.

Freeman, Derek. 1957. *The Family System of the Iban of Borneo.* Canberra: Department of Anthropology and Sociology, Australian National University.

Freeman, Derek. 1970. *Report on the Iban.* London: Athlone Press.

Fried, Stephanie Gorson. 2000. "Tropical Forests Forever? A Contextual Ecology of Bentian Rattan Agroforestry Systems." In *People, Plants, and Justice: The Politics of Nature Conservation*, edited by Charles Zerner. New York: Columbia University Press.

Fuentes, Agustin, and Linda D. Wolfe. 2002. *Primates Face to Face: Conservation Implications of Human-Nonhuman Primate Interconnections.* Cambridge: Cambridge University Press.

Galdikas, Biruté. 1988. "Orangutan Diet, Range, and Activity at Tanjung Puting, Central Borneo." *International Journal of Primatology* 9 (1): 1–35. doi: 10.1007/BF02740195.

Galdikas, Biruté M. F. 1985. "Orangutan Sociality at Tanjung Puting." *American Journal of Primatology* 9 (2): 101–19.

Galdikas, Biruté M. F. 1995. *Reflections of Eden: My Years with the Orangutans of Borneo.* Boston: Little, Brown.

Galdikas, Biruté M. F., and J. M. Wood. 1990. "Birth Spacing Patterns in Humans and Apes." *American Journal of Primatology* 6 (1): 49–51.

Geertz, Hildred. 1961. *The Javanese Family: A Study of Kinship and Socialization.* New York: Free Press.
George, Kenneth M. 1996. *Showing Signs of Violence: The Cultural Politics of a Twentieth-Century Headhunting Ritual.* Berkeley: University of California Press.
Gershon, Ilana. 2010. *The Breakup 2.0: Disconnecting over New Media.* Ithaca, NY: Cornell University Press.
Geschiere, Peter. 2009. *The Perils of Belonging: Autochthony, Citizenship, and Exclusion in Africa and Europe.* Chicago: University of Chicago Press.
Gibson-Graham, J. K. 2014. "Being the Revolution, or, How to Live in a 'More-Than--Capitalist' World Threatened with Extinction." *Rethinking Marxism* 26 (1): 76–94.
Gilmore, Ruth Wilson. 2007. *Golden Gulag: Prisons, Surplus, Crisis, and Opposition in Globalizing California.* Berkeley: University of California Press.
Gilroy, Paul. 1993. *The Black Atlantic: Modernity and Double Consciousness.* London: Verso.
Goeman, Mishuana. 2013. *Mark My Words: Native Women Mapping our Nations.* Minneapolis: University of Minnesota Press.
Goeman, Mishuana, and Jennifer Nez Denetdale. 2009. "Native Feminisms: Legacies, Interventions, and Indigenous Sovereignties." *Wicazo Sa Review* 24 (2): 9–13.
Good, Byron J., and Mary-Jo DelVecchio Good. 2001. "Why Do the Masses So Easily Run Amok?" *Latitudes* 5:10–19.
Goossens, B., L. Chikhi, Mohd. Fairus Jalil, Sheena James, Marc Ancrenaz, Isabelle Lackman-Ancrenaz, and Michael W. Bruford. 2009. "Taxonomy, Geographic Variation and Population Genetics of Bornean and Sumatran Orangutans." In *Orangutans: Geographic Variation in Behavioral Ecology and Conservation*, edited by S. Wich, S. Suci Utami Atmoko, Tatang Mitra Setia, and Carel P. van Schaik, 1–13. Oxford: Oxford University Press.
Goris, M. G., V. Kikken, M. Straetemans, S. Alba, M. Goeijenbier, E. C. van Gorp, K. R. Boer, J. F. Wagenaar, and R. A. Hartskeerl. 2013. "Towards the Burden of Human Leptospirosis: Duration of Acute Illness and Occurrence of Post--Leptospirosis Symptoms of Patients in the Netherlands." *PLoS ONE* 8 (10): e76549. doi: 10.1371/journal.pone.0076549.
Govindrajan, Radhika. 2015. "'The Goat That Died for Family': Animal Sacrifice and Interspecies Kinship in India's Central Himalayas." *American Ethnologist* 42 (3): 504–19. doi: 10.1111/amet.12144.
Grandia, Liza. 2012. *Enclosed: Conservation, Cattle, and Commerce among the Q'eqchi' Maya Lowlanders.* Seattle: University of Washington Press.
Green, R., and D. S. Mankikar. 1949. "Afebrile Cases of Melioidosis." *British Medical Journal* 1 (4598): 308–11. doi: 10.2307/25371309.
Greenhalgh, Susan. 2013. "Patriarchal Demographics? China's Sex Ratio Reconsidered." *Population and Development Review* 38:130–49. doi: 10.1111/j.1728-4457.2013.00556.x.
Greer, Jed, and Kenny Bruno. 1996. *Greenwash: The Reality behind Corporate Environmentalism.* Penang, Malaysia: Third World Network.

Grijpstra, B. J. 1976. *Common Efforts in the Development of Rural Sarawak, Malaysia*. Assen, Netherlands: Van Gorcum.
Grosz, Elizabeth. 1998. "The Time of Violence: Deconstruction and Value." *Cultural Values* 2 (2/3): 190–205.
Gruen, Lori. 2011. *Ethics and Animals: An Introduction*. Cambridge: Cambridge University Press.
Guan anak Sureng. 1960. "Six Orang Stories." *Sarawak Museum Journal* 9 (15–16): 6.
Hainsworth, D. R. 1992. *Stewards, Lords, and People: The Estate Steward and His World in Later Stuart England*. Cambridge: Cambridge University Press.
Hall, Jacquelyn Dowd. 1983. "The Mind That Burns in Each Body." In *Powers of Desire: The Politics of Sexuality*, edited by Ann Snitow, Christine Stansell, and Sharon Thompson, 328–49. New York: Monthly Review.
Hall, Stuart. 1996. "When Was 'the Post-Colonial'? Thinking at the Limit." In *The Post-Colonial Question: Common Skies, Divided Horizons*, edited by Iain Chambers and Lidia Curti, 242–60. London: Routledge.
Haraway, Donna. 2014. "Anthropocene, Capitalocene, Chthulucene: Staying with the Trouble." Presentation in Anthropocene: Arts of Living on a Damaged Planet, University of California–Santa Cruz, May 9, 2014. http://opentranscripts.org/transcript/anthropocene-capitalocene-chthulucene/.
Haraway, Donna J. 1988. "Situated Knowledges: The Science Question and the Privilege of Partial Perspective." *Feminist Studies* 14 (3): 575–99.
Haraway, Donna J. 1989. *Primate Visions: Gender, Race, and Nature in the World of Modern Science*. New York: Routledge.
Haraway, Donna J. 1991. *Simians, Cyborgs, and Women: The Reinvention of Nature*. New York: Routledge.
Haraway, Donna J. 2003. *The Companion Species Manifesto: Dogs, People, and Significant Otherness*. Chicago: Prickly Paradigm Press.
Haraway, Donna J. 2008. *When Species Meet*. Minneapolis: University of Minnesota Press.
Haraway, Donna J. 2016. *Staying with the Trouble: Making Kin in the Chthulucene*. Durham, NC: Duke University Press.
Harding, Andrew J., and James Chin, eds. 2014. *50 Years of Malaysia: Federalism Revisited*. Singapore: Marshall Cavendish.
Hardt, Michael. 1999. "Affective Labor." *Boundary 2* 26 (2): 89–100.
Hardt, Michael. 2007. "Foreward: What Affects Are Good For." In *The Affective Turn*, edited by Patricia Clough. Durham, NC: Duke University Press.
Hardt, Michael, and Antonio Negri. 2004. *Multitude: War and Democracy in the Age of Empire*. New York: Penguin.
Haritaworn, Jinthana, Adi Kuntsman, and Silvia Posocco. 2014. *Queer Necropolitics*. Abingdon: Routledge.
Harrison, Faye Venetia, ed. 1991. *Decolonizing Anthropology: Moving Further toward an Anthropology for Liberation*. Washington, DC: Association of Black Anthropologists, American Anthropological Association.

Harrisson, Barbara. 1963. "The Education of Young Orang-Utans to Living in the Wild." *Oryx* 7 (2–3): 2–3.
Harrisson, Barbara. 1965a. "Conservation Needs of the Orangutan." Proceedings of the Conference on Conservation of Nature and Natural Resources in Tropical South East Asia, Bangkok, Thailand, November 29–December 4.
Harrisson, Barbara. 1965b. "OURS: The Orang Utan Recovery Service 1963/65." Proceedings of the Conference on Conservation of Nature and Natural Resources in Tropical Southeast Asia, Bangkok, Thailand, November 29–December 4.
Harrisson, Barbara. [1962] 1987. *Orang-Utan*. Oxford: Oxford University Press.
Harrisson, Tom. 1959. *A World Within: A Borneo Story*. London: Cresset Press.
Harshey, Radhika, and Jonathan D. Partridge. 2015. "Shelter in a Swarm." *Journal of Molecular Biology* 427 (23): 3683–94.
Hartigan, John, Ron Eglash, Clarence C. Gravlee, Linda M. Hunt, et. al. *Anthropology of Race: Genes, Biology, and Culture*. Santa Fe: SAR Press, 2013.
Hartigan, John. n.d. "Flagellar Motors and the Work of Culture." In *How Nature Works*, edited by Sarah Besky and Alex Blanchette. Unpublished manuscript, last modified June 25, 2017. Microsoft Word file.
Hauser, M. D., D. Weiss, and G. Marcus. 2002. "RETRACTED: Rule Learning by Cotton-Top Tamarins." *Cognition* 86 (1): B15–B22. doi: 10.1016/s0010-0277(02)00139-7.
Hayden, Cori. 2003. *When Nature Goes Public: The Making and Unmaking of Bioprospecting in Mexico*. Princeton, NJ: Princeton University Press.
Hayward, E. V. A. 2010. "Fingeryeyes: Impressions of Cup Corals." *Cultural Anthropology* 25 (4): 577–99. doi: 10.1111/j.1548-1360.2010.01070.x.
He, Baogang, Brian Galligan, and Takashi Inoguchi. 2007. *Federalism in Asia*. Cheltenham, UK: Edward Elgar.
Heimann, Judith M. 1998. *The Most Offending Soul Alive: Tom Harrisson and His Remarkable Life*. Honolulu: University of Hawai'i Press.
Heise, Ursula K. 2016. *Imagining Extinction: The Cultural Meanings of Endangered Species*. Chicago: University of Chicago Press.
Helmreich, Stefan. 2009. *Alien Ocean: Anthropological Voyages in Microbial Seas*. Berkeley: University of California Press.
Herzig, Rebecca M. 2005. *Suffering for Science: Reason and Sacrifice in Modern America*. Piscataway, NJ: Rutgers University Press.
Ho, Engseng. 2006. *The Graves of Tarim: Genealogy and Mobility across the Indian Ocean*. Berkeley: University of California Press.
Ho, Karen Z. 2009. *Liquidated: An Ethnography of Wall Street*. Durham, NC: Duke University Press.
Hochschild, Arlie Russell. 1983. *The Managed Heart: Commercialization of Human Feeling*. Berkeley: University of California Press.
Hochschild, Arlie Russell. [1989] 2003. *The Second Shift*. New York: Penguin.
Hong, Evelyne. 1987. *Natives of Sarawak: Survival in Borneo's Vanishing Forest*. Penang: Instittute Masyarakat.

Horowitz, Leah Sophie. 1998. "Integrating Indigenous Resource Management with Wildlife Conservation: A Case Study of Batang Ai." *Human Ecology: An Interdisciplinary Journal* 26 (3): 371–403.

Hose, Charles. 1926. *Natural Man: A Record from Borneo.* London: Macmillan.

Hose, Charles, and W. McDougall. 1901. *The Relations between Men and Animals in Sarawak.* London: Anthropological Institute of Great Britain and Ireland.

Hughes, David McDermott. 2006a. *From Enslavement to Environmentalism: Politics on a Southern African Frontier.* Seattle: University of Washington Press.

Hughes, David McDermott. 2006b. "Hydrology of Hope: Farm Dams, Conservation, and Whiteness in Zimbabwe." *American Ethnologist* 33 (2): 269–87. doi: 10.1525/ae.2006.33.2.269.

Humboldt, Alexander von. 1868. *Cosmos: A Sketch of a Physical Description of the Universe.* New York: Harper and Brothers.

Hume, Joseph. 1853. *A Letter to the Earl of Malmesbury Relative to the Proceedings of Sir James Brooke in Borneo.* London.

Husson, S. J., D. Kurniawan, P. Purnomo, N. Boyd, A. Syuyoko, J. Sunderland--Groves, and J. Sihite. 2016. "The Survival and Adaptation of Reintroduced Ex-Captive Orangutans in Central Kalimantan, Indonesia." Proceedings of the International Primatological Society and American Society for Primatology Joint Meeting, Chicago, Illinois, August 22, 2016.

Husson, Simon J., Serge A. Wich, Andrew J. Marshall, Rona D. Dennis, Marc Ancrenaz, Rebecca Brassley, Melvin Gumal, Andrew J. Hearn, Erik Meijaard, Togu Simorangkir, and Ian Singleton. 2009. "Orangutan Distribution, Density, Abundance and Impacts of Disturbance." In *Orangutans: Geographical Variation in Behavioral Ecology and Conservation*, edited by Serge A. Wich, S. Suci Utami Atmoko, Tatang Mitra Setia, and Carel P. van Schaik. Oxford: Oxford University Press.

Ibrahim, Nur Amali. 2018. *Improvising Islam: Indonesian Youths in a Time of Possibility.* Ithaca, NY: Cornell University Press.

Ibrahim, Yasamin Kh, Lim Tze Tshen, Kira E. Westaway, Earl of Cranbrook, Louise Humphrey, Ros Fatihah Muhammad, Jian-xin Zhao, and Lee Chai Peng. 2013. "First Discovery of Pleistocene Orangutan (*Pongo* sp.) Fossils in Peninsular Malaysia: Biogeographic and Paleoenvironmental Implications." *Journal of Human Evolution* 65 (6): 770–97.

Idrus, Rusaslina. 2010. "From Wards to Citizens: Indigenous Rights and Citizenship in Malaysia." *Political and Legal Anthropology Review* 33 (1): 89–108. doi: 10.1111/j.1555-2934.2010.01094.x.

Idrus, Rusaslina. 2016. "Reading Che Guevara in Kuala Lumpur: The Rise of Independent, Political, and Intellectual Youth Collectives in Malaysia." Annual Conference of the Association for Asian Studies, Seattle, Washington, April 2.

Ileto, Reynaldo. 1992. "Religion and Anti-Colonial Movements." In *The Cambridge History of Southeast Asia*, edited by Nicholas Tarling, 197–248. Cambridge: Cambridge University Press.

Ingold, Tim. 2007. *Lines: A Brief History.* London: Routledge.

IUCN. 2016. "IUCN Red List of Threatened Species." Accessed October 27, 2016. http://www.iucnredlist.org.

Jacobs, Harriet A., Lydia Maria Child, and Jean Fagan Yellin. 1987. *Incidents in the Life of a Slave Girl: Written by Herself.* Cambridge, MA: Harvard University Press.

Jain, S. Lochlann. 2013. *Malignant: How Cancer Becomes Us.* Berkeley: University of California Press.

Jawan, Jayum A. 1991. *The Ethnic Factor in Modern Politics: The Case of Sarawak, East Malaysia.* Hull, UK: Center for South-East Asian Studies, University of Hull.

Jawan, Jayum A. 1993. *The Iban Factor in Sarawak Politics.* Serdang, Malaysia: University Pertanian Malaysia Press.

Jawan, Jayum A. 1994. *Iban Politics and Economic Development: Their Patterns and Change.* Bangi, Malaysia: Penerbit University Kebangsaan Malaysia.

Jeffrey, Julie Roy. 1998. *Frontier Women: "Civilizing" the West? 1840–1880*, rev. ed. New York: Hill and Wang.

Jensen, Erik. 1974. *The Iban and Their Religion.* Oxford: Clarendon Press.

Johnson, Robert. 2014. "Kant's Moral Philosophy." http://plato.stanford.edu/archives/sum2014/entries/kant-moral/. Accessed December 15, 2016.

Johnson, Walter. 2013. *River of Dark Dreams : Slavery and Empire in the Cotton Kingdom.* Cambridge, MA: Belknap Press.

Kapoor, Ilan. 2012. *Celebrity Humanitarianism: The Ideology of Global Charity.* New York: Routledge.

Kauanui, J. Kehaulani. 2008. *Hawaiian Blood: Colonialism and the Politics of Sovereignty and Indigeneity.* Durham, NC: Duke University Press.

Kaur, Amarjit. 1998. "A History of Forestry in Sarawak." *Modern Asian Studies* 32 (1): 117–47. doi:10.1017/S0026749X98003011.

Keating, Christine. 2011. *Decolonizing Democracy: Transforming the Social Contract in India.* University Park: Pennsylvania State University Press.

Kedit, Peter Mulok. 1980. *Modernization among the Iban of Sarawak.* Kuala Lumpur: Dewan Bahasa dan Pustaka, Kementerian Pelajaran, Malaysia.

Keller, Arthur, Walther Birk, and Axel Tagesson Möller. 1911. *Kinderpflege--Lehrbuch.* Berlin: J. Springer.

Keller, Evelyn Fox. 1982. "Feminism and Science." *Signs* 7 (3): 589–602.

Khalili, Laleh. 2013. *Time in the Shadows: Confinement in Counterinsurgencies.* Stanford, CA: Stanford University Press.

King, Barbara. 2013. "Why Gorillas Aren't Sexist and Orangutans Don't Rape." National Public Radio. Accessed October 3, 2013. http://www.npr.org/sections/13.7/2013/10/03/228809153/why-gorillas-arent-sexist-and-orangutans-dont-rape.

King, Victor T. 1986. "Planning for Agrarian Change: Hydro-Electric Power, Resettlement and Iban Swidden Cultivators in Sarawak, East Malaysia." Occasional Papers. Hull, UK: Centre for South-East Asian Studies, University of Hull.

King, Victor T. 1993. *The Peoples of Borneo.* The Peoples of South-East Asia and the Pacific. Oxford: Blackwell.

Kipling, Rudyard. 1919. *Two Tales*. Boston: International Pocket Library.
Kirksey, Eben. 2012. *Freedom in Entangled Worlds: West Papua and the Architecture of Global Power*. Durham, NC: Duke University Press.
Kirksey, S. Eben, and Stefan Helmreich. 2010. "The Emergence of Multispecies Ethnography." *Cultural Anthropology* 25 (4): 545–76.
Kleinman, Arthur, Veena Das, and Margaret M. Lock. 1997. *Social Suffering*. Berkeley: University of California Press.
Klima, Alan. 2002. *The Funeral Casino: Meditation, Massacre, and Exchange with the Dead in Thailand*. Princeton, NJ: Princeton University Press.
Knight, Andrew. 2008. "The Beginning of the End for Chimpanzee Experiments?" *Philosophy, Ethics, and Humanities in Medicine* 3 (1): 1–14.
Knott, C. D., M. E. Thompson, R. M. Stumpf, and M. H. McIntyre. 2010. "Female Reproductive Strategies in Orangutans: Evidence for Female Choice and Counterstrategies to Infanticide in a Species with Frequent Sexual Coercion." *Proceedings of the Royal Society B—Biological Sciences* 277 (1678): 105–13. doi: 10.1098/rspb.2009.1552.
Knott, Cheryl, Lydia Beaudrot, Tamaini Snaith, Sarah White, Hartmut Tschauner, and George Planansky. 2008. "Female-Female Competition in Bornean Orangutans." *International Journal of Primatology* 29 (4): 975–97.
Kohn, Eduardo. 2007. "How Dogs Dream: Amazonian Natures and the Politics of Transspecies Engagement." *American Ethnologist* 34 (1): 3–24. doi: 10.1525/ae.2007.34.1.3.
Kohn, Eduardo. 2013. *How Forests Think: Towards an Anthropology beyond the Human*. Berkeley: University of California Press.
Kolbert, Elizabeth. 2014. *The Sixth Extinction: An Unnatural History*. New York: Henry Holt and Company.
Komer, Robert. 1972. *The Malayan Emergency in Retrospect: Organization of a Successful Counterinsurgency Effort*. Santa Monica, CA: RAND.
Kon, Y. 1994. "Amok." *British Journal of Psychiatry* 165:685–89. doi: 10.1192/bjp.165.5.685.
Kornhauser, Ryan, and Kathy Laster. 2014. "Punitiveness in Australia: Electronic Monitoring vs the Prison." *Crime, Law and Social Change* 62 (4): 445–74. doi: 10.1007/s10611-014-9538-2.
Kosek, Jake. 2006. *Understories: The Political Life of Forests in Northern New Mexico*. Durham, NC: Duke University Press.
Koselleck, Reinhart. 1985. *Futures Past: On the Semantics of Historical Time*. Cambridge, MA: MIT Press.
Kua Kia, Soong. 2008. "Racial Conflict in Malaysia: Against the Official History." *Race and Class* 49 (3): 33–53.
Kuze, Noko, David Dellatore, Graham L. Banes, Peter Pratje, Tomoyuki Tajima, and Anne E. Russon. 2012. "Factors Affecting Reproduction in Rehabilitant Female Orangutans: Young Age at First Birth and Short Inter-Birth Interval." *Primates* 53 (2): 181–92.
Kuze, Noko, Symphorosa Sipangkui, Titol Peter Malim, Henry Bernard, Laurentius

N. Ambu, and Shiro Kohshima. 2008. "Reproductive Parameters over a 37-Year Period of Free-Ranging Female Borneo Orangutans at Sepilok Orangutan Rehabilitation Centre." *Primates* 49 (2): 126–34. doi: 10.1007/s10329-008-0080-7.

Langford, Jean M. 2013. "Wilder Powers: Morality and Animality in Tales of War and Terror." *Hau: Journal of Ethnographic Theory* 3 (3): 223–44.

Langford, Jean M. 2017. "Avian Bedlam." *Environmental Humanities* 9 (1): 84–107.

Latour, Bruno. 1993. *We Have Never Been Modern*. Translated by Catherine Porter. Cambridge, MA: Harvard University Press.

Latour, Bruno, and Peter Weibel, eds. 2005. *Making Things Public: Atmospheres of Democracy*. Cambridge, MA: MIT Press.

Latour, Bruno, and Steve Woolgar. 1979. *Laboratory Life: The Social Construction of Scientific Facts*. Beverly Hills, CA: SAGE.

Lee, Doreen. 2016. *Activist Archives: Youth Culture and the Political Past in Indonesia*. Durham, NC: Duke University Press.

Lee, Jia Hui. n.d. "'Accept Them As They Are': *Lelaki Lembut* as a Category of Effeminacy in Malaysia." Unpublished manuscript, last modified December 3, 2013. PDF.

Lee, Y. H., Y. Chen, X. Ouyang, and Y. H. Gan. 2010. "Identification of Tomato Plant as a Novel Host Model for *Burkholderia pseudomallei*." *BMC Microbiology* 10:28. doi: 10.1186/1471-2180-10-28.

Le Guin, Ursula K. 2014. "Ursula K. Le Guin's Speech at National Book Awards: 'Books Aren't Just Commodities.'" *The Guardian*. http://www.theguardian.com/books/2014/nov/20/ursula-k-le-guin-national-book-awards-speech. Accessed November 20, 2014.

Leigh, Michael B. 1974. *The Rising Moon: Political Change in Sarawak*. Sydney: Sydney University Press.

Lenin, Janaki. 2013. "The Rape Issue." *The Hindu*, April 5. http://www.thehindu.com/features/metroplus/society/the-rape-issue/article4584578.ece. Accessed April 20, 2013.

Leonie, Amanda, Jenifer Lasimbang, Holly Jonas, and Benedict Mansul. 2015. *Red and Raw: Indigenous Rights in Malaysia and the Law: A Research Report on the Legal Framework of Indigenous Peoples Rights in Malaysia*. Kuala Lumpur: Jaringan Orang Asal Semalaysia.

Lévi-Strauss, Claude. 1969. *The Elementary Structures of Kinship*, rev. ed. Edited by Rodney Needham. Translated by James H. Bell. Boston: Beacon Press.

Li, Tania. 2007. *The Will to Improve: Governmentality, Development, and the Practice of Politics*. Durham, NC: Duke University Press.

Liebal, Katja, Cornelia Müller, and Simone Pika. 2007. *Gestural Communication in Nonhuman and Human Primates*. Amsterdam: J. Benjamins.

Liebal, Katja, Simone Pika, and Michael Tomasello. 2006. "Gestural Communication of Orangutans (*Pongo pygmaeus*)." *Gesture* 6 (1): 1–38.

Lindee, Susan, and Joanna Radin. 2016. "Patrons of the Human Experience: A History of the Wenner-Gren Foundation for Anthropological Research, 1941–2016." *Current Anthropology* 57 (Sup. 14): 218–301.

Livingston, Julie. 2012. *Improvising Medicine: An African Oncology Ward in an Emerging Cancer Epidemic*. Durham, NC: Duke University Press.

Livingston, Julie, and Jasbir K. Puar. 2011. "Interspecies." *Social Text* 29 (1): 3–13.

Lockard, Craig A., and Graham E. Saunders. 1972. *Old Sarawak: A Pictorial Study*. Kuching, Malaysia: Borneo Literature Bureau.

Longrich, N. R., J. Scriberas, and M. A. Wills. 2016. "Severe Extinction and Rapid Recovery of Mammals across the Cretaceous-Palaeogene Boundary, and the Effects of Rarity on Patterns of Extinction and Recovery." *Journal of Evolutionary Biology* 29 (8): 1495–1512. doi: 10.1111/jeb.12882.

Lorimer, Jamie. 2015. *Wildlife in the Anthropocene: Conservation after Nature*. Minneapolis: University of Minnesota Press.

Lowe, Celia. 2006. *Wild Profusion: Biodiversity Conservation in an Indonesian Archipelago*. Princeton, NJ: Princeton University Press.

Lugones, Maria. 2010. "Toward a Decolonial Feminism." *Hypatia: A Journal of Feminist Philosophy* 25 (4): 742–59.

Lyons, Kristina. 2016. "Decomposition as Life Politics: Soils, Selva, and Small Farmers under the Gun of the U.S.-Colombian War on Drugs." *Cultural Anthropology* 31 (1): 56–81.

M'Charek, Amade. 2013. "Beyond Fact or Fiction: On the Materiality of Race in Practice." *Cultural Anthropology* 28 (3): 420–42.

Mackenzie, Catriona, and Natalie Stoljar, eds. 2000. *Relational Autonomy: Feminist Perspectives on Autonomy, Agency, and the Social Self*. New York: Oxford University Press.

MacKinnon, John. 1974. "The Behaviour and Ecology of Wild Orangutans (*Pongo Pygmaeus*)." *Animal Behaviour* 22 (1): 3–74.

Maggioncalda, Anne N., Nancy M. Czekala, and Robert M. Sapolsky. 2002. "Male Orangutan Subadulthood: A New Twist on the Relationship between Chronic Stress and Developmental Arrest." *American Journal of Physical Anthropology* 118 (1): 25–32.

Mahavera, Sheridan. 2010. "How Poor Are We, Really?" *Malaysian Insider*. http://www.themalaysianinsider.com/malaysia/article/how-poor-are-we-really. Accessed May 2, 2011.

Majumder, Atreyee. 2013. "Calling Out to the Faraway: Accessing History through Public Gestures in Howrah, West Bengal." *South Asia Multidisciplinary Academic Journal* 7. http://samaj.revues.org/3614.

Mamdani, Mahmood. 1996. *Citizen and Subject: Contemporary Africa and the Legacy of Late Colonialism*. Princeton, NJ: Princeton University Press.

Mamdani, Mahmood. 2001. *When Victims Become Killers: Colonialism, Nativism, and the Genocide in Rwanda*. Princeton, NJ: Princeton University Press.

Mankekar, Purnima, and Akhil Gupta. 2016. "Intimate Encounters: Affective Labor in Call Centers." *Positions: Asia Critique* 24 (1): 17–43.

Mansfield, Becky, Christine Biermann, Kendra McSweeney, Justine Law, Caleb Gallemore, Leslie Horner, and Darla K. Munroe. 2014. "Environmental Politics after Nature: Conflicting Socioecological Futures." *Annals of the Associa-*

tion of American Geographers 105 (2): 284–93. doi: 10.1080/00045608.2014.973802.
Margulis, Lynn. 1981. *Symbiosis in Cell Evolution: Life and Its Environment on the Early Earth.* San Francisco: W. H. Freeman.
Marshall, A. J., Nardiyono, L. M. Engström, B. Pamungkas, E. Meijaard, and S. A. Stanley. 2006. "The Blowgun Is Mighter Than the Chainsaw in Determining Population Density of Bornean Orangutans (*Pongo pygmaeus morio*) in the Forests of East Kalimantan." *Biological Conservation* 129:566–78.
Martin, Aryn, Natasha Myers, and Ana Viseu. 2015. "The Politics of Care in Technoscience." *Social Studies of Science* 45 (5): 625–41.
Marx, Karl. 1981. *Capital: A Critique of Political Economy.* New York: Penguin.
Masco, Joseph. 2006. *The Nuclear Borderlands: The Manhattan Project in Post–Cold War New Mexico.* Princeton, NJ: Princeton University Press.
Massey, Doreen B. 1994. *Space, Place, and Gender.* Minneapolis: University of Minnesota Press.
Massey, Doreen B. 2005. *For Space.* London: SAGE.
Massumi, Brian. 2002. *Parables for the Virtual: Movement, Affect, Sensation.* Durham NC: Duke University Press.
Mbembe, J. A. 2001. *On the Postcolony.* Berkeley: University of California Press.
McClintock, Anne. 1995. *Imperial Leather: Race, Gender, and Sexuality in the Colonial Contest.* New York: Routledge.
McDow, Thomas. 2018. *Buying Time: Debt and Mobility in the Indian Ocean World.* Athens: Ohio University Press.
McGehee, Nancy Gard. 2014. "Volunteer Tourism: Evolution, Issues and Futures." *Journal of Sustainable Tourism* 22 (6): 847–54.
McGranahan, Carole. 2010. *Arrested Histories: Tibet, the CIA, and Memories of a Forgotten War.* Durham, NC: Duke University Press.
M'Charek, Amade. 2013. "Beyond Fact or Fiction: On the Materiality of Race in Practice." *Cultural Anthropology* 28 (3):420–42.
Meijaard, E., D. Buchori, Y. Hadiprakarsa, S. S. Utami-Atmoko, A. Nurcahyo, A. Tjiu, D. Prasetyo, Nardiyono, L. Christie, M. Ancrenaz, F. Abadi, I. N. Antoni, D. Armayadi, A. Dinato, Ella, P. Gumelar, T. P. Indrawan, Kussaritano, C. Munajat, C. W. Priyono, Y. Purwanto, D. Puspitasari, M. S. Putra, A. Rahmat, H. Ramadani, J. Sammy, D. Siswanto, M. Syamsuri, N. Andayani, H. Wu, J. A. Wells, and K. Mengersen. 2011. "Quantifying Killing of Orangutans and Human-Orangutan Conflict in Kalimantan, Indonesia." *PloS ONE* 6 (11): e27491.
Meijaard, Erik, Serge Wich, Marc Ancrenaz, and Andrew J. Marshall. 2012. "Not by Science Alone: Why Orangutan Conservationists Must Think outside the Box." *Annals of the New York Academy of Sciences* 1249 (1): 29–44.
Merchant, Carolyn. 1980. *The Death of Nature: Women, Ecology, and the Scientific Revolution.* San Francisco: HarperSanFrancisco.
Merry, Sally Engle. 2016. *The Seductions of Quantification: Measuring Human Rights, Gender Violence, and Sex Trafficking.* Chicago: University of Chicago Press.
Middleton, C. Townsend. 2011. "Across the Interface of State Ethnography: Re-

thinking Ethnology and Its Subjects in Multicultural India." *American Ethnologist* 38 (2): 249–66. doi: 10.1111/j.1548-1425.2011.01304.x.
Midgley, Clare. 1992. *Women against Slavery: The British Campaigns, 1780–1870*. London: Routledge.
Mignolo, Walter. 2000. *Local Histories/Global Designs: Coloniality, Subaltern Knowledges, and Border Thinking*. Princeton, NJ: Princeton University Press.
Mignolo, Walter. 2015. "Sylvia Wynter: What Does It Mean to Be Human?" In *Sylvia Wynter: On Being Human as Praxis*, edited by Katherine McKittrick. Durham, NC: Duke University Press.
Miles, H. Lyn White. 1993. "Language and the Orang-Utan: The Old 'Person' of the Forest." In *The Great Ape Project: Equality beyond Humanity*, edited by Paola Cavalieri and Peter Singer. New York: St. Martin's Griffin.
Mills, Sara. 1993. *Discourses of Difference: An Analysis of Women's Travel Writing and Colonialism*. London: Routledge.
Milner, Anthony. 1994. *The Invention of Politics in Colonial Malaya*. Cambridge: University of Cambridge Press.
Mitani, John C. 1985. "Mating Behaviour of Male Orangutans in the Kutai Game Reserve, Indonesia." *Animal Behaviour* 33 (2): 392–402. doi: 10.1016/s0003-3472(85)80063-4.
Mitchell, Timothy. 2002. *Rule of Experts: Egypt, Techno-Politics, Modernity*. Berkeley: University of California Press.
Mittermeier, Russell A., Norman Myers, Jorgen B. Thomsen, Gustavo A. B. da Fonseca, and Silvio Olivieri. 1998. "Biodiversity Hotspots and Major Tropical Wilderness Areas: Approaches to Setting Conservation Priorities." *Conservation Biology* 12 (3): 516–20. doi: 10.1046/j.1523-1739.1998.012003516.x.
Mol, Annemarie. 2002. *The Body Multiple: Ontology in Medical Practice, Science and Cultural Theory*. Durham, NC: Duke University Press.
Mor, Vincent, and Susan Allen. 2006. "Hospice." In *The Encyclopedia of Aging*, edited by Richard Schulz, Linda S. Noelke, Kenneth Rockwood, and Richard L. Sprott. New York: Springer.
Morrogh-Bernard, H. C., S. J. Husson, S. E. Page, and J. O. Rieley. 2003. "Population Status of the Bornean Orangutan (*Pongo pygmaeus*) in the Sebangau Peat Swamp Forest, Central Kalimantan, Indonesia." *Biological Conservation* 110:141–52.
Mortimer-Sandilands, Catriona, and Bruce Erickson. 2010. *Queer Ecologies: Sex, Nature, Politics, Desire*. Bloomington: Indiana University Press.
Muehlebach, Andrea. 2011. "On Affective Labor in Post-Fordist Italy." *Cultural Anthropology* 26 (1): 59–82. doi: 10.1111/j.1548-1360.2010.01080.x.
Muehlebach, Andrea Karin. 2012. *The Moral Neoliberal: Welfare and Citizenship in Italy*. Chicago: University of Chicago Press.
Muehlenbein, Michael P., Leigh Ann Martinez, Andrea A. Lemke, Laurentius Ambu, Senthilvel Nathan, Sylvia Alsisto, and Rosman Sakong. 2010. "Unhealthy Travelers Present Challenges to Sustainable Primate Ecotourism." *Travel Medicine and Infectious Disease* 8 (3): 169–75. doi: http://dx.doi.org/10.1016/j.tmaid.2010.03.004.
Mukharji, Projit Bihari. 2014. "Vishalyakarani as *Eupatorium ayapana*: Retro--

botanizing, Embedded Traditions, and Multiple Historicities of Plants in Colonial Bengal, 1890–1940." *Journal of Asian Studies* 73 (1): 65–87. doi: 10.1017/S0021911813001733.
Muñoz, José Esteban. 2007. "Cruising the Toilet: LeRoi Jones/Amiri Baraka, Radical Black Traditions, and Queer Futurity." *GLQ: A Journal of Lesbian and Gay Studies* 13 (2): 353–67.
Murphy, Michelle. 2015. "Unsettling Care: Troubling Transnational Itineraries of Care in Feminist Health Practices." *Social Studies of Science* 45 (5): 717–37.
Murphy, Michelle. 2016. "Alterlife in the Ongoing Aftermath: Exposure, Entanglement, Survivance." Presentation at Toxic: A Symposium on Exposure, Entanglement, and Endurance, Yale University, March 1.
Murphy, Michelle. 2017. *The Economization of Life*. Durham, NC: Duke University Press.
Myers, Norman, and Russell A. Mittermeier. 2000. "Biodiversity Hotspots for Conservation Priorities." *Nature* 403 (6772): 853–58.
Nadler, R. D., and D. C. Collins. 1991. "Copulatory Frequency, Urinary Pregnanediol, and Fertility in Great Apes." *American Journal of Primatology* 24 (3–4): 167–79. doi: 10.1002/ajp.1350240304.
Nagata, Judith. 2011. "Boundaries of Malayness: "We Have Made Malaysia: Now It Is Time to (Re)Make the Malays but Who Interprets the History?" In *Melayu: The Politics, Poetics, and Paradoxes of Malayness*, edited by Maznah Mohamad and Syed Muhd Khairudin Aljunied, 3–33. Singapore: National University of Singapore Press.
Nelson, Diane M. 2015. *Who Counts? The Mathematics of Death and Life after Genocide*. Durham, NC: Duke University Press.
Ngai, Sianne. 2012. *Our Aesthetic Categories: Zany, Cute, Interesting*. Cambridge, MA: Harvard University Press.
Ngũgĩ wa, Thiong'o. 1986. *Decolonising the Mind: The Politics of Language in African Literature*. London: J. Currey.
Nihon Kankyō Kaigi. 2009. *The State of the Environment in Asia 2006/2007*. Tokyo: United Nations University Press.
Nixon, Rob. 2011. *Slow Violence and the Environmentalism of the Poor*. Cambridge, MA: Harvard University Press.
Obeyesekere, Gananath. 1992. *The Apotheosis of Captain Cook: European Mythmaking in the Pacific*. Princeton, NJ: Princeton University Press.
Oliver-Smith, Anthony, ed. 2009. *Development and Dispossession: The Crisis of Forced Displacement and Resettlement*. Santa Fe, NM: School for Advanced Research Press.
Omi, Michael, and Howard Winant. [1986] 2015. *Racial Formation in the United States*, 3rd ed. New York: Routledge.
O'Neill, Kevin Lewis. 2013. "Left Behind: Security, Salvation, and the Subject of Prevention." *Cultural Anthropology* 28 (2): 204–26.
Ong, Aihwa. 1987. *Spirits of Resistance and Capitalist Discipline: Factory Women in Malaysia*. Albany: State University of New York Press.

Ong, Aihwa. 1990. "State versus Islam: Malay Families, Women's Bodies, and the Body Politic in Malaysia." *American Ethnologist* 17 (2): 258–76.
Ong, Kee Hui. 1998. *Footprints in Sarawak: Memoirs of Tan Sri Datuk (Dr) Ong Kee Hui*, 2 vols. Kuching, Malaysia: Research and Resource Centre.
Ongkili, James P. 1967. *The Borneo Response to Malaysia, 1961–1963*. Brisbane: Donald Moore.
Padoch, Christine. 1982. *Migration and Its Alternatives among the Iban of Sarawak*. The Hague: Nijhoff.
Padoch, Christine, and Nancy Lee Peluso. 1996. *Borneo in Transition: People, Forests, Conservation, and Development*. Kuala Lumpur: Oxford University Press.
Palmer, Alexandra, Nicholas Malone, and Julie Park. 2016. "Caregiver/Orangutan Relationships at Auckland Zoo: Empathy, Friendship, and Ethics between Species." *Society and Animals* 24 (3): 230–49.
Pandian, Anand. 2008. "Pastoral Power in the Postcolony: On the Biopolitics of the Criminal in South India." *Cultural Anthropology* 23 (1): 85–117.
Parreñas, Juno Salazar. 2016. "The Materiality of Intimacy: Rethinking 'Ethical Capitalism' through Embodied Encounters with Animals in Southeast Asia." *Positions: Asia Critique* 24 (1): 97–127.
Parreñas, Rhacel Salazar. 2001a. "Mothering from a Distance." *Feminist Studies* 27 (2): 361–90.
Parreñas, Rhacel Salazar. 2001b. *Servants of Globalization: Women, Migration, and Domestic Work*. Stanford, CA: Stanford University Press.
Parreñas, Rheana "Juno" Salazar. 2012. "Producing Affect: Transnational Volunteerism in a Malaysian Orangutan Rehabilitation Center." *American Ethnologist* 39 (4): 673–87. doi: 10.1111/j.1548-1425.2012.01387.x.
Pashal anak Dagang, Julia anak Sang, Malcom anak Demies, Bibian M. Diway, Oswald Braken Tisen, and Lucy Chong. 2013. *WiMOR: Wildlife Monitoring and Rescue Operation at Bakun Hydro Electric Project Flooded Zone*, edited by Katherine G. Pearce. Kuching, Malaysia: Sarawak Forestry Corporation.
Paxson, Heather. 2013. *The Life of Cheese: Crafting Food and Value in America*. Berkeley: University of California Press.
Payne, Junaidi, and J. Cede Prudente. 2008. *Orangutans: Behavior, Ecology, and Conservation*. Cambridge, MA: MIT Press.
Payne, Robert. [1960] 1986. *The White Rajahs of Sarawak*. Singapore: Oxford University Press.
Peletz, Michael G. 1996. *Reason and Passion: Representations of Gender in a Malay Society*. Berkeley: University of California Press.
Peluso, Nancy Lee. 1991. *Rich Forests, Poor People: Resource Control and Resistance in Java*. Berkeley: University of California Press.
Peluso, Nancy Lee, and Christian Lund. 2011. "New Frontiers of Land Control: Introduction." *Journal of Peasant Studies* 38 (4): 667–81. doi: 10.1080/03066150.2011.607692.
Peluso, Nancy Lee, and Michael Watts. 2001. *Violent Environments*. Ithaca, NY: Cornell University Press.

Petryna, Adriana. 2002. *Life Exposed: Biological Citizens after Chernobyl*. Princeton, NJ: Princeton University Press.

Phillips, John. 2006. "Agencement/Assemblage." *Theory, Culture and Society* 23 (2–3): 108–9. doi: 10.1177/026327640602300219.

Pick, Daniel. 1989. *Faces of Degeneration: A European Disorder, c. 1848–c. 1918*. Cambridge: Cambridge University Press.

Podin, Y., D. S. Sarovich, E. P. Price, M. Kaestli, M Mayo, K. Hii, H. Ngian, S. Wong, I. Wong, J. Wong, A. Mohan, M. Ooi, T. Fam, J. Wong, A. Tuanyok, P. Keim, P. M. Giffard, and B. J. Currie. 2014. "*Burkholderia pseudomallei* Isolates from Sarawak, Malaysian Borneo, Are Predominantly Susceptible to Aminoglycosides and Macrolides." *Antimicrobial Agents and Chemotherapy* 58 (1): 162–66.

Polanyi, Karl. 2001. *The Great Transformation: The Political and Economic Origins of Our Time*. Boston, MA: Beacon Press.

Porritt, Vernon L. 1997. *British Colonial Rule in Sarawak, 1946–1963*. South-East Asian Historical Monographs. Kuala Lampur: Oxford University Press.

Porter, Natalie. 2013. "Bird Flu Biopower: Strategies for Multispecies Coexistence in Việt Nam." *American Ethnologist* 40 (1): 132–48.

Povinelli, Daniel J., Laura A. Theall, James E. Reaux, and Sarah Dunphy-Lelii. 2003. "Chimpanzees Spontaneously Alter the Location of Their Gestures to Match the Attentional Orientation of Others." *Animal Behaviour* 66:71–79. doi: 10.1006/anbe.2003.2195.

Povinelli, Elizabeth A. 2001. "Radical Worlds: The Anthropology of Incommensurability and Inconceivability." *Annual Review of Anthropology* 30: 319–34. doi: 10.2307/3069219.

Povinelli, Elizabeth A. 2006. *The Empire of Love: Toward a Theory of Intimacy, Genealogy, and Carnality*. Durham, NC: Duke University Press.

Pratt, Mary Louise. 1992. *Imperial Eyes: Travel Writing and Transculturation*. London: Routledge.

Priedhorsky, Reid, Jilin Chen, Shyong (Tony) K. Lam, Katherine Panciera, Loren Terveen, and John Riedl. 2007. "Creating, Destroying, and Restoring Value in Wikipedia." GROUP '07 Proceedings of the 2007 International ACM Conference on Supporting Group Work, Sanibel Island, FL, November 4–7.

Pringle, Robert. 1970. *Rajahs and Rebels: The Ibans of Sarawak under Brooke Rule, 1841–1941*. London: Macmillan.

Quijano, Aníbal. 1995. "Modernity, Identity, and Utopia in Latin America." In *The Postmodernism Debate in Latin America*, edited by John Beverley, Michael Aronna, and Jose Oviedo. Durham, NC: Duke University Press.

Rafael, Vicente. 2000. *White Love: And Other Events in Filipino History*. Durham, NC: Duke University Press.

Raffles, Hugh. 2002. *In Amazonia: A Natural History*. Princeton, NJ: Princeton University Press.

Raffles, Hugh. 2010. *Insectopedia*. New York: Pantheon.

"Rajah Agrees to Cession of Sarawak." 1946. *Morning Bulletin* (Rockhampton,

Queensland), February 8. Accessed August 24, 2015. http://nla.gov.au/nla.news-article56433754.

Ramirez, Catherine S. 2008. "Afrofuturism/Chicanafuturism: Fictive Kin." *Aztlan: A Journal of Chicano Studies* 33 (1): 185–94.

Reddy, Sagili Chandrasekhara. 2012. "Ocular Injuries by Durian Fruit." *International Journal of Ophthalmology* 5 (4): 530–34. doi: 10.3980/j.issn.2222-3959.2012.04.25.

Reece, Bob. 1982. *The Name of Brooke: The End of White Rajah Rule in Sarawak.* Kuala Lumpur: Oxford University Press.

Reger, Jo. 2014. "Micro-Cohorts, Feminist Discourse, and the Emergence of the Toronto SlutWalk." *Feminist Formations* 26 (1): 49–69.

Reid, Anthony. 1998. "Merdeka: The Concept of Freedom in Indonesia." In *Asian Freedoms: The Idea of Freedom in East and Southeast Asia*, edited by David Kelly and Anthony Reid, 141–60. Cambridge: Cambridge University Press.

Reid, Anthony. 2000. *Charting the Shape of Early Modern Southeast Asia.* Singapore: Institute of Southeast Asian Studies.

Reyes, Herman. 2007. "Force-Feeding and Coercion: No Physician Complicity." *Virtual Mentor: American Medical Association Journal of Ethics* 9 (10): 904–8.

Rich, Adrienne Cecile. 1981. *Compulsory Heterosexuality and Lesbian Existence.* London: Onlywomen.

Riley, Erin P. 2007. "The Human–Macaque Interface: Conservation Implications of Current and Future Overlap and Conflict in Lore Lindu National Park, Sulawesi, Indonesia." *American Anthropologist* 109 (3): 473–84. doi: 10.1525/aa.2007.109.3.473.

Ritzer, George. 2015. *The McDonaldization of Society*, 8th ed. Los Angeles: SAGE.

Robbins, Joel, Bambi B. Schieffelin, and Aparecida Vilaça. 2014. "Evangelical Conversion and the Transformation of the Self in Amazonia and Melanesia: Christianity and the Revival of Anthropological Comparison." *Comparative Studies in Society and History* 56 (3): 559–90.

Rodman, Peter S., and John G. H. Cant. 1984. *Adaptations for Foraging in Nonhuman Primates: Contributions to an Organismal Biology of Prosimians, Monkeys, and Apes.* New York: Columbia University Press.

Roelvink, Gerda, Kevin St. Martin, and J. K. Gibson-Graham. 2015. *Making Other Worlds Possible: Performing Diverse Economies.* Minneapolis: University of Minnesota Press.

Root-Bernstein, Meredith, Leo Douglas, A. Smith, and Diogo Verissimo. 2013. "Anthropomorphized Species as Tools for Conservation: Utility beyond Prosocial, Intelligent and Suffering Species."*Biodiversity and Conservation* 22 (8): 1577–89.

Root-Bernstein, Robert, Jessica Vonck, and Abigail Podufaly. 2009. "Antigenic complementarity between coxsackie virus and streptococcus in the induction of rheumatic heart disease and autoimmune myocarditis." *Autoimmunity* 42 (1): 1–16.

Rosaldo, Renato. 1980. *Ilongot Headhunting, 1883–1974: A Study in Society and History.* Stanford, CA: Stanford University Press.

Rosaldo, Renato. 1989. "Imperialist Nostalgia." *Representations* 26:107–22.
Rosaldo, Renato. 2013. *The Day of Shelly's Death: The Poetry and Ethnography of Grief.* Durham, NC: Duke University Press.
Rose, Deborah Bird. 2004. *Reports from a Wild Country: Ethics for Decolonisation.* Sydney: University of New South Wales Press.
Rosenberg, Daniel, and Susan Harding. 2005. *Histories of the Future.* Durham, NC: Duke University Press.
Rosenberg, Gabriel N. 2016. "A Race Suicide among the Hogs: The Biopolitics of Pork in the United States, 1865–1930." *American Quarterly* 68 (1): 49–73.
Rothfels, Nigel. 2002. *Savages and Beasts: The Birth of the Modern Zoo.* Baltimore: Johns Hopkins University Press.
Rubin, Gayle. 1975. "The Traffic in Women: Notes on the 'Political Economy' of Sex." In *Toward an Anthropology of Women*, edited by Rayna Reiter. New York: Monthly Review.
Russ, Ann Julienne. 2005. "Love's Labor Paid For: Gift and Commodity at the Threshold of Death." *Cultural Anthropology* 20 (1): 128–55.
Russon, A., and K. Andrews. 2010. "Orangutan Pantomime: Elaborating the Message." *Biology Letters* 7 (4): 627–30. doi: 10.1098/rsbl.2010.0564.
Russon, A. E., A. Erman, and R. Dennis. 2001. "The Population and Distribution of Orangutans (*Pongo pygmaeus pygmaeus*) in and around the Danau Sentarum Wildlife Reserve, West Kalimantan, Indonesia." *Biological Conservation* 97:21–28.
Russon, Anne E. 1999. *Orangutans: Wizards of the Rain Forest.* Toronto: Key Porter.
Rutherford, Danilyn. 2012. *Laughing at Leviathan: Sovereignty and Audience in West Papua.* Chicago: University of Chicago Press.
Rutherford, Danilyn. 2016. "Affect Theory and the Empirical." *Annual Review of Anthropology* 45 (1): 285–300.
Sahlins, Marshall. 1995. *How "Natives" Think: About Captain Cook, for Example.* Chicago: University of Chicago Press.
Sahlins, Marshall. 1999. "What Is Anthropological Enlightenment? Some Lessons of the Twentieth Century." *Annual Review of Anthropology* 28:i–xxiii. doi: 10.1146/annurev.anthro.28.1.0.
Said, Edward W. 1993. *Culture and Imperialism.* New York: Knopf.
Sanday, Peggy Reeves. 1981. "The Socio-Cultural Context of Rape: A Cross--Cultural Study." *Journal of Social Issues* 37 (4): 5–27. doi: 10.1111/j.1540-4560.1981.tb01068.x.
Sandin, Benedict. 1967. *The Sea Dayaks of Borneo: Before White Rajah Rule.* London: Macmillan.
Sandin, Benedict. 1980. *Iban Adat and Augury.* Penang, Malaysia: Penerbit University Sains Malaysia for Social Comparative Studies.
Sarawak Museum. 1979. *Batang Ai Hydro-Electric Project: Survey on the Attitudes of the Affected People towards the Project and Resettlement.* Kuching, Malaysia: Sarawak Museum.
Sarawak Museum Curator. 1965. "Orang-Utan from Kalimantan." Official memorandum, Sarawak Museum, archival file MU/444/1.

Schlegel, Friedrich. [1800–1801] 1996. "Philosophical Lectures: Transcendental Philosophy." In *The Early Political Writings of the German Romantics*, edited by Frederick C. Beiser. Cambridge: Cambridge University Press.

Schmidt, K., L. Hill, and G. Guthrie. 1977. "Running Amok." *International Journal of Social Psychiatry* 23 (4): 264–74. doi: 10.1177/002076407702300405.

Schrader, Astrid. 2015. "Abyssal Intimacies and Temporalities of Care: How (Not) to Care about Deformed Leaf Bugs in the Aftermath of Chernobyl." *Social Studies of Science* 45 (5): 665–90.

Schrader, Astrid. 2017. "Microbes." In *Gender: Animals*, edited by Juno Salazar Parreñas. Farmington Hills, MI: Macmillan Reference USA.

Scott, James C. 1998. *Seeing Like a State.* New Haven, CT: Yale University Press.

Scott, James C. 2009. *The Art of Not Being Governed: An Anarchist History of Upland Southeast Asia*. New Haven, CT: Yale University Press.

Scott, James C. 2017. *Against the Grain: A Deep History of the Earliest States.* New Haven, CT: Yale University Press.

Seymour, Nicole. 2013. *Strange Natures: Futurity, Empathy, and the Queer Ecological Imagination*. Urbana: University of Illinois Press.

Shapiro, Nicholas. 2015. "Attuning to the Chemosphere: Domestic Formaldehyde, Bodily Reasoning, and the Chemical Sublime." *Cultural Anthropology* 30 (3): 368–93. doi: 10.14506/ca30.3.02.

Sharma, Aradhana. 2006. "Crossbreeding Institutions, Breeding Struggle: Women's Empowerment, Neoliberal Governmentality, and State (Re)Formation in India." *Cultural Anthropology* 21 (1): 60–95. doi: 10.1525/can.2006.21.1.60.

Sharpe, Christina Elizabeth. 2016. *In the Wake: On Blackness and Being.* Durham. NC: Duke University Press.

Sheehy, Thomas W., John J. Deller, and David R. Weber. 1967. "Melioidosis." *Annals of Internal Medicine* 69 (4): 897–900. http://proxy.lib.ohio-state.edu/login?url=http://search.ebscohost.com/login.aspx?direct=true&db=a9h&AN=6966509&site=ehost-live.

Shellabear, Rev. W. G. 1916. *An English-Malay Dictionary*. Singapore: Methodist Publishing House.

Shryock, Andrew, Daniel Lord Smail, and Timothy K. Earle. 2011. *Deep History: The Architecture of Past and Present.* Berkeley: University of California Press.

Sibon, Peter. 2011. "No Mystery in Deaths." *Borneo Post*, August 3. Accessed November 1, 2011. http://www.theborneopost.com/2011/08/03/no-mystery-in-deaths/.

Simpson, Audra. 2014. *Mohawk Interruptus: Political Life across the Borders of Settler States*. Durham, NC: Duke Universitiy Press.

Singer, Peter. 1975. *Animal Liberation: A New Ethics for Our Treatment of Animals.* New York: New York Review.

Singleton, Ian, Cheryl. D. Knott, Helen C. Morrogh-Bernard, Serge A. Wich, and Carel P. van Schaik. 2010. "Ranging Behavior of Orangutan Females and Social Organization." In *Orangutans: Geographic Variation in Behavioral Ecology and Conservation*, edited by Serge A. Wich, S. Suci Utami Atmoko, Tatang Mitra Setia, and Carel P. van Shaik. Oxford: Oxford University Press.

Sivaramakrishnan, K. 1999. *Modern Forests: Statemaking and Environmental Change in Colonial Eastern India*. Stanford, CA: Stanford University Press.
Smith, Andrea. 2010. "Queer Theory and Native Studies: The Heteronormativity of Settler Colonialism." *GLQ: A Journal of Lesbian and Gay Studies* 16 (1–2): 41–68. doi: 10.1215/10642684-2009-012.
Smith, Linda Tuhiwai. 1999. *Decolonizing Methodologies: Research and Indigenous Peoples*. London: Zed.
Smuts, Barbara. 1987. "Sexual Competition and Mate Choice." In *Primate Societies*, edited by Barbara Smuts, Dorothy Cheney, Robert Seyfarth, Richard W. Wrangham, and Thomas Struhsaker, 385–99. Chicago: University of Chicago Press.
Smythies, B. 1963. "History of Forestry in Sarawak." *Malayan Forester* 26:232–52.
Sodikoff, Genese. 2009. "The Low-Wage Conservationist: Biodiversity and Perversities of Value in Madagascar." *American Anthropologist* 111 (4): 443–55. doi: 10.1111/j.1548-1433.2009.01154.x.
Sodikoff, Genese Marie. 2012. *The Anthropology of Extinction: Essays on Culture and Species Death*. Bloomington: Indiana University Press.
Solomon, Daniel. 2016. "Interpellation and Affect: Activating Political Potentials across Primate Species at Jakhoo Mandir, Shimla." *Humanimalia: A Journal of Human/Animal Interface Studies* 8 (1): 1–34.
South-East Asia Treaty Organization. 1973. *The Communist Insurgency in Sarawak*. Short Paper no. 57. Bangkok: Research Office, South-East Asia Treaty Organization.
Spivak, Gayatri Chakravorty. 1988. "Can the Subaltern Speak?" In *Marxism and the Interpretation of Culture*, edited by Cary Nelson and Lawrence Grossberg. Urbana: University of Illinois Press.
Spivak, Gayatri Chakravorty. 1999. *A Critique of Postcolonial Reason: Toward a History of the Vanishing Present*. Cambridge, MA: Harvard University Press.
Spock, Benjamin. 1946. *The Common Sense Book of Baby and Child Care*. New York: Duell, Sloan and Pearce.
Spock, Benjamin, and Elke Kaspar. 1952. *Dein Kind—dein Glück*. Stuttgart: Hatje.
Spores, John C. 1988. *Running Amok: An Historical Inquiry*. Athens: Ohio University Center for International Studies.
Sprague, L. D., and H. Neubauer. 2004. "Melioidosis in Animals: A Review on Epizootiology, Diagnosis and Clinical Presentation." *Journal of Veterinary Medicine B, Infectious Diseases and Veterinary Public Health* 51 (7): 305–20.
Spyer, Patricia. 2000. *The Memory of Trade: Modernity's Entanglements on an Eastern Indonesian Island*. Durham, NC: Duke University Press.
Steedly, Mary Margaret. 1993. *Hanging without a Rope: Narrative Experience in Colonial and Postcolonial Karoland*. Princeton, NJ: Princeton University Press.
Steedly, Mary Margaret. 2013. *Rifle Reports: A Story of Indonesian Independence*. Berkeley: University of California Press.
Stevenson, Lisa. 2012. "The Psychic Life of Biopolitics: Survival, Cooperation, and Inuit Community." *American Ethnologist* 39 (3): 592–613. doi: 10.1111/j.1548-1425.2012.01383.x.

Stoler, Ann Laura. 1995. *Race and the Education of Desire*. Durham, NC: Duke University Press.
Stoler, Ann Laura. 2002. *Carnal Knowledge and Imperial Power: Race and the Intimate in Colonial Rule*. Berkeley: University of California Press.
Stoler, Ann Laura. 2013. *Imperial Debris: On Ruins and Ruination*. Durham, NC: Duke University Press.
Stoler, Ann Laura, and Karen Strassler. 2000. "Castings for the Colonial: Memory Work in 'New Order' Java." *Comparative Studies in Society and History* 42 (1): 4–48.
Stoler, Ann Laura, and Karen Strassler. 2002. "Memory-Work in Java: A Cautionary Tale." In *Carnal Knowledge and Imperial Power: Race and the Intimate in Colonial Rule*, by Ann Laura Stoler. Berkeley: University of California Press.
Stumpf, R. M., M. E. Thompson, and C. D. Knott. 2008. "A Comparison of Female Mating Strategies in *Pan troglodytes* and *Pongo* spp." *International Journal of Primatology* 29 (4): 865–84. doi: 10.1007/s10764-008-9284-3.
Subramaniam, Banu. 2014. *Ghost Stories for Darwin: The Science of Variation and the Politics of Diversity*. Urbana: University of Illinois Press.
Subramanian, Ajantha. 2009. *Shorelines: Space and Rights in South India*. Stanford, CA: Stanford University Press.
Tagliacozzo, Eric. 2005. *Secret Trades, Porous Borders: Smuggling and States along a Southeast Asian Frontier, 1865–1915*. New Haven, CT: Yale University Press.
Taib Mahmud, and Gaya Media Sdn. Bhd. 1996. *Taib: Wira pembangunan Sarawak = Hero of Development*. Kuching, Malaysia: Gaya Media.
TallBear, Kimberly. 2013. *Native American DNA: Tribal Belonging and the False Promise of Genetic Science*. Minneapolis: University of Minnesota Press.
Tamarkin, Noah. 2011. "Religion as Race, Recognition as Democracy: Lemba Black Jews in South Africa." *Annals of the American Academy of Political and Social Science* 637 (1): 148–64.
Taussig, Michael T. 1980. *The Devil and Commodity Fetishism in South America*. Chapel Hill: University of North Carolina Press.
Tawie, Joseph. 2011. "Mysterious Deaths in Bakun Dam Area." *Free Malaysia Today*, August 1. http://www.freemalaysiatoday.com/category/nation/2011/08/01/mysterious-deaths-in-bakun-dam-area/. Accessed August 5, 2011.
Tempelmann, Sebastian, Juliane Kaminski, and Katja Liebal. 2011. "Focus on the Essential: All Great Apes Know When Others Are Being Attentive." *Animal Cognition* 14 (3): 433–39. doi: 10.1007/s10071-011-0378-5.
Theidon, Kimberly Susan. 2013. *Intimate Enemies: Violence and Reconciliation in Peru*. Philadelphia: University of Pennsylvania Press.
Thomas, Deborah A. 2011. *Exceptional Violence: Embodied Citizenship in Transnational Jamaica*. Durham, NC: Duke University Press.
Thomas, Keith. 1996. *Man and the Natural World: Changing Attitudes in England, 1500–1800*. New York: Oxford University Press.
Thompson, Charis. 2005. *Making Parents: The Ontological Choreography of Reproductive Technologies*. Cambridge, MA: MIT Press.

Thompson, E. P. 1968. *The Making of the English Working Class.* Harmondsworth, UK: Penguin.
Thompson, E. P. 1975. *Whigs and Hunters: The Origin of the Black Act.* New York: Pantheon.
Tonry, Michael. 2001. "Symbol, Substance, and Severity in Western Penal Policies." *Punishment and Society* 3 (4): 517–36.
Trouillot, Michel-Rolph. 1995. *Silencing the Past: Power and the Production of History.* Boston: Beacon Press.
Tsing, Anna. 2003. "Cultivating the Wild: Honey-Hunting and Forest Management in Southeast Kalimantan." In *Culture and the Question of Rights: Forests, Coasts, and Seas in Southeast Asia*, edited by Charles Zerner, 24–55. Durham, NC: Duke University Press.
Tsing, Anna. 2005. *Friction: An Ethnography of Global Connection.* Princeton, NJ: Princeton University Press.
Tsing, Anna. 2015. *The Mushroom at the End of the World: On the Possibility of Life in Capitalist Ruins.* Princeton, NJ: Princeton University Press.
Tsing, Anna Lowenhaupt. 1993. *In the Realm of the Diamond Queen: Marginality in an Out-of-the-Way Place.* Princeton, NJ: Princeton University Press.
Tuan, Yi-fu. 1979. *Landscapes of Fear.* Minneapolis: University of Minnesota Press.
Tuan, Yi-fu. 1984. *Dominance and Affection: The Making of Pets.* New Haven, CT: Yale University Press.
Tuck, Eve, and K. Wayne Yang. 2012. "Decolonization Is Not a Metaphor." *Decolonization: Indigeneity, Education and Society* 1 (1): 1–40.
Uddin, Lisa. 2015. *Zoo Renewal: White Flight and the Animal Ghetto.* Minneapolis: University of Minnesota Press.
van Dooren, Thom. 2014. *Flight Ways: Life and Loss at the Edge of Extinction.* New York: Columbia University Press.
van Rossum. Matthias. 2015. *Kleurrijke tragiek: De geschiedenis van slavernij in Azië onder de VOC.* Hilversum, Netherlands: Uitgeverij Verloren.
van Schaik, C. P., M. A. van Noordwijk, and E. R. Vogel. 2009. "Ecological Sex Differences in Wild Orangutans." In *Orangutans: Geographic Variation in Behavioral Ecology and Conservation*, edited by Serge A. Wich, S. Suci Utami Atmoko, Tatang Mitra Setia, T., and Carel P. van Schaik. Oxford: Oxford University Press.
Varela, Francisco. 1979. *Principles of Biological Autonomy.* New York: Elsevier North Holland.
Vergara, Camilo José. 1995. *The New American Ghetto.* Piscataway, NJ: Rutgers University Press.
Vergara, Camilo José. 1999. *American Ruins.* New York: Monacelli.
Vink, Markus. 2003. "'The World's Oldest Trade': Dutch Slavery and Slave Trade in the Indian Ocean in the Seventeenth Century." *Journal of World History* 14 (2): 131–77.
Voigt, Maria, Serge A. Wich, Marc Ancrenaz, et al. 2018. "Global Demand for Natural Resources Eliminated More Than 100,000 Bornean Orangutans." *Current Biology* 28 (1-9). https://doi.org/10.1016/j.cub.2018.01.053.

von Oertzen, Christine, Maria Rentetzi, and Elizabeth S. Watkins. 2013. "Finding Science in Surprising Places: Gender and the Geography of Scientific Knowledge. Introduction to 'Beyond the Academy: Histories of Gender and Knowledge.'" *Centaurus* 55 (2): 73–80.

Wadley, Reed L., and Carol J. Pierce Colfer. 1997. "Hunting Primates and Managing Forests: The Case of Iban Forest Farmers in Indonesian Borneo." *Human Ecology: An Interdisciplinary Journal* 25 (2): 243–71.

Wagner, Ulla. 1972. *Colonialism and Iban Warfare.* Stockholm: OBE-Tryck.

Wainwright, Joel. 2008. *Decolonizing Development: Colonial Power and the Maya.* Malden, MA: Blackwell.

Wallace, Alfred Russel. [1869] 1986. *The Malay Archipelago: The Land of the Orang-Utan, and the Bird of Paradise.* Singapore: Oxford University Press.

Wallerstein, Immanuel Maurice. 1974. *The Modern World-System.* New York: Academic Press.

Ward, Arthur Bartlett. 1966. *Rajah's Servant.* Ithaca, NY: Cornell University Press.

Warinner, Christina, and Cecil M. Lewis. 2015. "Microbiome and Health in Past and Present Human Populations." *American Anthropologist* 117 (4): 740–41.

Watanabe, Chika. 2015. "Commitments of Debt: Temporality and the Meanings of Aid Work in a Japanese NGO in Myanmar." *American Anthropologist* 117 (3): 460–79. doi: 10.1111/aman.12287.

Watts, D. P., M. Muller, S. J. Amsler, G. Mbabazi, and J. C. Mitani. 2006. "Lethal Intergroup Aggression by Chimpanzees in Kibale National Park, Uganda." *American Journal of Primatology* 68:161–80.

Weaver, Harlan. 2013. "'Becoming in Kind': Race, Class, Gender, and Nation in Cultures of Dog Rescue and Dogfighting." *American Quarterly* 65 (3): 689–709.

Weheliye, Alexander G. 2014. *Habeas Viscus: Racializing Assemblages, Biopolitics, and Black Feminist Theories of the Human.* Durham, NC: Duke University Press.

Weiss, Nancy Pottishman. 1977. "Mother, the Invention of Necessity: Dr. Benjamin Spock's Baby and Child Care." *American Quarterly* 29 (5): 519–46. doi: 10.2307/2712572.

Welker, Marina. 2014. *Enacting the Corporation: An American Mining Firm in Post-Authoritarian Indonesia.* Berkeley: University of California Press.

Wells-Barnett, Ida B. 1892. *Southern Horrors: Lynch Law in All Its Phases.* New York: New York Age.

West, Mary Mills. 1914. *Infant Care.* Washington, DC: U.S. Government Printing Office.

West, Paige. 2006. *Conservation Is Our Government Now: The Politics of Ecology in Papua New Guinea.* Durham, NC: Duke University Press.

White, Richard. 1995. *The Organic Machine.* New York: Hill and Wang.

Whitmore, A., and C. J. Krishnaswami. 1912. "An Account of a Hitherto Undescribed Infective Disease Occuring among the Population of Rangoon." *Indian Medical Gazette* 47:262.

Whorf, Benjamin Lee. 1956. *Language, Thought, and Reality: Selected Writings of Benjamin Lee Whorf.* Cambridge, MA: MIT Press.

Wich, Serge A. 2009. *Orangutans: Geographic Variation in Behavioral Ecology and Conservation*. Oxford: Oxford University Press.
Wich, Serge A., and H. Kuehl. 2016. "The Future of the Sumatran Orangutan (*Pongo abelii*)." International Primatological Society and American Society for Primatology Joint Meeting, Chicago, Illinois, August 24.
Wilder, Gary. 2015. *Freedom Time: Negritude, Decolonization, and the Future of the World*. Durham, NC: Duke University Press.
Williams, Mark, Jan Zalasiewicz, P. K. Haff, Christian Schwägerl, Anthony D. Barnosky, and Erle C. Ellis. 2015. "The Anthropocene Biosphere." *Anthropocene Review* 2 (3): 196–219. doi: 10.1177/2053019615591020.
Winichakul, Thongchai. 1994. *Siam Mapped: A History of the Geo-Body of a Nation*. Honolulu: University of Hawai'i Press.
Winkler, L. A. 1995. "A Brief Review of Studies of Orangutan Morphology and Development with a Discussion of Their Relevancy to Physical Anthropology." In *The Neglected Ape*, edited by R. D. Nadler, 251–66. New York: Plenum.
Woese, Carl R., Otto Kandler, and Mark L. Wheelis. 1990. "Towards a Natural System of Organisms: Proposal for the Domains Archaea, Bacteria, and Eucarya." *Proceedings of the National Academy of Sciences of the United States of America* 87 (12): 4576–79.
Wolfe, Cary. 2003. *Animal Rites: American Culture, the Discourse of Species, and Posthumanist Theory*. Chicago: University of Chicago Press.
"Woman Queried over '#Tangkap Najib' Sticker on Car." 2015. *Borneo Post*, August 12. http://www.theborneopost.com/2015/08/12/woman-queried-over-tangkap-najib-sticker-on-car/.
Wood, Bernard, and Terry Harrison. 2011. "The Evolutionary Context of the First Hominins." *Nature* 470 (7334): 347–52. doi: 10.1038/nature09709.
Wool, Zoë Hamilton. 2015. *After War: The Weight of Life at Walter Reed*. Durham, NC: Duke University Press.
World Commission on Dams. 1999. *Empty Promises, Damned Lives: Final Report of the Fact Finding Mission, 7th–14th May 1999: Evidence from the Bakun Resettlement Scheme in Sarawak*. Petaling Jaya: Suaram Komunikasi.
Wrangham, Richard W., and Dale Peterson. 1996. *Demonic Males: Apes and the Origins of Human Violence*. Boston: Houghton Mifflin.
Wright, Melissa W. 2011. "Necropolitics, Narcopolitics, and Femicide: Gendered Violence on the Mexico-U.S. Border." *Signs: Journal of Women in Culture and Society* 36 (3): 707–31.
Yanagisako, Sylvia Junko. 2002. *Producing Culture and Capital: Family Firms in Italy*. Princeton, NJ: Princeton University Press.
Yingst, Samuel L., Paul Facemire, Lara Chuvala, David Norwood, Mark Wolcott, and Deron A. Alves. 2014. "Pathological Findings and Diagnostic Implications of a Rhesus Macaque (*Macaca mulatta*) Model of Aerosol-Exposure Melioidosis (*Burkholderia pseudomallei*)." *Journal of Medical Microbiology* 63: 118–28.
Yong, Kee Howe. 2013. *The Hakkas of Sarawak: Sacrificial Gifts in Cold War Era Malaysia*. Toronto: University of Toronto Press.

Yong, Stephen K. T. 1998. *A Life Twice Lived: A Memoir.* Kuching, Malaysia: Author.

Young, Robin. 2015. "Students Can Live For Free at Dutch Nursing Home." *Here and Now.* Boston: WBUR. Aired April 7.

Zain, Nizam, 2014. "Pasaran terbuka menindas rakyat, zalim." *Sinar Harian.* http://www.sinarharian.com.my/mobile/nasional/pasaran-terbuka-menindas-rakyat-zalim-1.254219.

Zaloom, Caitlin. 2006. *Out of the Pits: Traders and Technology from Chicago to London.* Chicago: University of Chicago Press.

Zee, Jerry C. 2017. "Holding Patterns: Sand and Political Time at China's Desert Shores." *Cultural Anthropology* 32 (2): 215–41.

Zerner, Charles. 2000. *People, Plants, and Justice: The Politics of Nature Conservation.* New York: Columbia University Press.

Žižek, Slavoj. 2010. *Living in the End Times.* London: Verso.

Zuk, Marlene. 2002. *Sexual Selections: What We Can and Can't Learn about Sex from Animals.* Berkeley: University of California Press.

INDEX

Page numbers followed by *f* indicate an illustration.